Communications in Computer and Information Science 3027

Rationale

The CCIS series is devoted to the publication of proceedings of computer science conferences. Its aim is to efficiently disseminate original research results in informatics in printed and electronic form. While the focus is on publication of peer-reviewed full papers presenting mature work, inclusion of reviewed short papers reporting on work in progress is welcome, too. Besides globally relevant meetings with internationally representative program committees guaranteeing a strict peer-reviewing and paper selection process, conferences run by societies or of high regional or national relevance are also considered for publication.

Topics

The topical scope of CCIS spans the entire spectrum of informatics ranging from foundational topics in the theory of computing to information and communications science and technology and a broad variety of interdisciplinary application fields.

Information for Volume Editors and Authors

Publication in CCIS is free of charge. No royalties are paid, however, we offer registered conference participants temporary free access to the online version of the conference proceedings on SpringerLink (http://link.springer.com) by means of an http referrer from the conference website and/or a number of complimentary printed copies, as specified in the official acceptance email of the event.

CCIS proceedings can be published in time for distribution at conferences or as post-proceedings, and delivered in the form of printed books and/or electronically as USBs and/or e-content licenses for accessing proceedings at SpringerLink. Furthermore, CCIS proceedings are included in the CCIS electronic book series hosted in the SpringerLink digital library at http://link.springer.com/bookseries/7899. Conferences publishing in CCIS are allowed to use our online conference service (Meteor) for managing the whole proceedings lifecycle (from submission and reviewing to preparing for publication) free of charge.

Publication process

The language of publication is exclusively English. Authors publishing in CCIS have to sign the Springer CCIS copyright transfer form, however, they are free to use their material published in CCIS for substantially changed, more elaborate subsequent publications elsewhere. For the preparation of the camera-ready papers/files, authors have to strictly adhere to the Springer CCIS Authors' Instructions and are strongly encouraged to use the CCIS LaTeX style files or templates.

Abstracting/Indexing

CCIS is abstracted/indexed in DBLP, Google Scholar, EI-Compendex, Mathematical Reviews, SCImago, Scopus. CCIS volumes are also submitted for the inclusion in ISI Proceedings.

How to start

To start the evaluation of your proposal for inclusion in the CCIS series, please send an e-mail to ccis@springer.com

Nina Merkert · Alexander Herzog
Editors

Simulation Science

Third International Workshop, SimScience 2025
Clausthal-Zellerfeld, Germany, May 7–9, 2025
Revised Selected Papers

Springer

Editors
Nina Merkert
TU Clausthal
Clausthal-Zellerfeld, Germany

Alexander Herzog
Simulation Science Center
Clausthal-Göttingen
Clausthal-Zellerfeld, Germany

ISSN 1865-0929 ISSN 1865-0937 (electronic)
Communications in Computer and Information Science
ISBN 978-3-032-29719-8 ISBN 978-3-032-29720-4 (eBook)
https://doi.org/10.1007/978-3-032-29720-4

This Springer imprint is published by the registered company Springer Nature Switzerland AG
The registered company address is: Gewerbestrasse 11, 6330 Cham, Switzerland

Preface

This volume contains the papers from the Third Clausthal-Göttingen International Workshop on Simulation Science in Clausthal-Zellerfeld. Simulation-based analysis and engineering techniques are traditionally a research focus of Clausthal University of Technology and Georg-August University of Göttingen, which is especially reflected in their common interdisciplinary research cluster "Simulation Science Center Clausthal-Göttingen". The third "Clausthal-Göttingen International Workshop on Simulation Science" aimed to bring together researchers and practitioners from both industry and academia to report on the latest advances in simulation science. Artificial Intelligence is a powerful tool that in many cases allows more rapid and more efficient evaluation and prediction of research data compared to conventional methods. However, the predictive power of artificial intelligence depends on the availability of large data sets. Therefore, there are various scenarios in different disciplines including natural science, materials engineering, economics and social sciences in which simulations are needed. The workshop therefore focused on both simulations and AI and their interaction.

The conference covered the topics material science and engineering, socio-technical systems and optimization of networks. The selected papers come from researchers from Germany, France and Niger. The 7 full papers presented were carefully reviewed and selected from 44 submissions.

All accepted papers were peer reviewed in a single-blind process by three qualified reviewers based on their qualifications and experience.

The proceedings editors wish to thank the dedicated Technical Program Committee members and all the other reviewers for their contributions. The submission, registration and payments were managed by VDE conference services. We also thank Springer for their trust and for publishing the proceedings of SimScience 2025.

February 2026

Nina Merkert
Alexander Herzog

Organization

Joint Organizing Committee

Nina Merkert	TU Clausthal, Germany
Robert Bredereck	TU Clausthal, Germany
Gunther Brenner	TU Clausthal, Germany
Robert Mettin	Georg-August University of Göttingen, Germany

Program Committee

Eduardo M. Bringa	Universidad de Mendoza, Argentina
Sandrine Brochard	Université de Poitiers CNRS-ENSMA, France
Valentina Cacchiani	University of Bologna, Italy
Laura De Lorenzis	ETH Zürich, Switzerland
Umut Durak	DLR Braunschweig, Germany
Maximilian Merkert	TU Braunschweig, Germany
Helmut Neukirchen	University of Iceland, Iceland
Narayan Rangaraj	IIT Bombay, India
Ulrich Rieder	Ulm University, Germany
Yudi Rosandi	University of Padjadjaran, Indonesia
Carlos J. Ruestes	IMDEA Materials Institute, Spain
Siegfried Schmauder	University of Stuttgart, Germany
Rüdiger Schwarze	TU Freiberg, Germany
Nimrod Talmon	Ben-Gurion University of the Negev, Israel
Adri van Duin	Penn State University, USA
Roberto Gomes de Aguiar Veiga	University of São Paulo, Brazil
Giuseppe Vizzari	University of Milano-Bicocca, Italy
Sigrid Wenzel	University of Kassel, Germany
Gregory Zacharewicz	IMT Mines Alès, France
Zhibo Zhang	Guangdong Academy of Sciences, China
Horst Zisgen	Darmstadt University of Applied Sciences, Germany

Project Support

Masoom Shaban	TU Clausthal, Germany

Contents

New Insights into Iron's Dynamic Response through Molecular Dynamics Simulations

Nourou Amadou[1(✉)] and Thibaut de Rességuier[2]

[1] Département de Physique, Université Abdou Moumouni de Niamey, BP 10662 Niamey, Niger
nourou.amadou@uam.edu.ne

[2] Institut Pprime, CNRS, ENSMA, Université de Poitiers, Poitiers, France

Abstract. Understanding the response of iron to dynamic loading is a major scientific issue of interest for various application fields. Physical processes governing the dynamic behavior of iron at very high strain rates are highly coupled, from elastic response, plastic deformation, polymorphic (solid-solid) transformations, to melting, damage, spall failure and fragmentation. Characterizing these processes and understanding their coupling at the atomic scale require heavy and complex experiments, usually carried out at large-scale facilities, as well as Molecular Dynamics (MD) simulations, which now give access to temporal and spatial scales comparable to those met in experiments. In this proceeding paper, we report on new insights into iron's dynamic response through MD simulations. Specifically, we will discuss a post-yield hardening-like effect that shifts the bcc to hcp phase transition to unexpectedly high pressures, as well as release and recompression waves emitted from the phase transition front. These results illustrate the capability of simulations to capture not only the details of wave profiles and deformation mechanisms at the atomic level but also new phenomena yet to be uncovered experimentally.

Keywords: Plasticity · Phase Transformation · Iron · Shock and Ramp loading · Release and Recompression waves · Molecular Dynamics Simulations

1 Introduction

Under dynamic compression, at moderate strain rates, iron is traditionally expected to exhibit three distinct wave regimes [1–4]. The first is an elastic precursor, which brings the material to the point of permanent deformation at approximately 1 GPa. The second is a plastic wave (referred to as the P1 wave), through which permanent deformation occurs, mediated by either twinning deformation or dislocation slip. This wave compresses the material to the boundary of the ambient body-centered cubic (bcc) phase stability at pressures around 13 GPa [1, 2]. The third is a phase transition wave (denoted as P2 wave), which transforms the bcc phase into the high-pressure hexagonal close-packed (hcp) phase and compresses this phase to the final pressure.

The study of these waves has been crucial in advancing the understanding of iron's behavior, and solid allotropic materials in general, under extreme conditions induced

N. Merkert and A. Herzog (Eds.): SimScience 2025, CCIS 3027, pp. 1–14, 2026.
https://doi.org/10.1007/978-3-032-29720-4_1

by either shock or ramp wave propagation. Indeed, this three-wave structure has been extensively investigated at both experimental and theoretical levels, where it is now established that the onset pressures for plasticity and phase transition are complex functions of loading rate, thermodynamic path, initial sample conditions, etc. [5–17].

Experimentally, the vast majority of dynamic compression studies use VISAR (Velocity Interferometer System for Any Reflector) to measure the free-surface or interface velocities and evidence these distinct waves propagating through the sample [1, 2, 9–17]. In addition, a limited number of studies provide direct insights into the underlying atomic-scale deformation mechanisms, employing advanced diagnostics such as X-ray diffraction, Extended X-ray Absorption Fine Structure (EXAFS), or X-ray Absorption Near-Edge Structure (XANES) [5–7, 18, 19].

On the theoretical side, early modeling efforts focused on developing kinetic models that, when coupled with hydrodynamic equations, are capable of reproducing VISAR measurements. These models incorporate both elastic-plastic yielding models, such as the Johnson and Rhode model [20, 21], a multiscale strength model comprising dislocation mobility and work hardening [12, 22], etc., and phase transition behaviors, using approaches such as the Hayes model [23, 24], Avrami-type kinetics [16, 17, 25, 26], and phase-field methods [27].

More recently, with the advent of high-performance computing and the availability of reliable interatomic potentials, non-equilibrium molecular dynamics (NEMD) simulations have emerged as powerful tools for capturing not only the details of the wave profiles but also the atomic-level deformation mechanisms and their coupling, thereby enhancing the interpretation of experimental results [28–30]. Furthermore, these simulations could also reveal phenomena that remain experimentally unexplored. Thus, in this proceedings paper, we summarize our recent work, in which new insights into iron's dynamic response were reported through molecular dynamics simulations. Specifically, we discuss a hardening-like effect that shifts the bcc to hcp phase transition to higher pressures than expected, as well as release and recompression waves emitted from the phase transition front. These recent results illustrate the capability of MD to capture detailed features of dynamic deformation yet to be uncovered experimentally.

2 Method and Computational Details

Non-Equilibrium Molecular Dynamics (NEMD) simulations were performed to investigate the response of iron to dynamic ramp compression. Samples with a size along the loading direction of $1400a_0$ (a_0 is the lattice constant) were simulated for times up to 80 ps using the LAMMPS Molecular Dynamics code [https://www.lammps.org, [31]. The interaction between atoms was modeled using the Embedded-Atom Method (EAM) formalism, where the total energy of a system of N atoms is determined by considering each atom to be embedded in an environment composed of the other N-1 atoms [32, 33]. The total energy is computed as:

$$E = \sum_{i=1}^{N} E_i \tag{1}$$

where:

$$E_i = \frac{1}{2}\sum_j \phi(r_{ij}) + F(\rho_i) \tag{2}$$

$\phi(r_{ij})$ represents the effective interaction potential between atoms i and j separated by a distance r_{ij}, and F_i is the embedding energy representing the average effect of the electrons (with density ρ_i) of the *N-1* other atoms on atom *i*. The iron model used is the modified version of the Ackland potential [34], which predicts the phase transition from bcc to hcp iron at 13.75 GPa under equilibrium hydrodynamic compression, in very good agreement with experimental data [1, 2, 4]. This potential has been successfully used in previous work to study plasticity and phase transition in polycrystalline [30, 35] iron under shock, as well as in Fe-C alloys under hydrodynamic compression [36].

The samples were initially equilibrated at zero pressure and 50 K within the NPT ensemble and then dynamically deformed using the NVE ensemble. The dynamic ramp compression was applied on the sample surface by driving an effective infinite-mass wall piston with an imposed velocity $U_p(t)$ along the *Z*-direction, while periodic boundary conditions were maintained in the transverse directions. Piston velocities ranging from 600 m/s to 1600 m/s were explored. Local thermodynamic and mechanical variables, such as longitudinal stress (P_z), particle velocity (V_z), and shear stress, defined as $\frac{1}{2}(P_z - \frac{1}{2}(P_x + P_y))$, were evaluated within a spatial planar bin (of $3a_0$ width) perpendicular to the shock-wave propagation direction. The sample's hydrodynamical evolution was studied by generating isocontour plots from these hydrodynamic parameters.

Furthermore, local phases and defects were analyzed using adaptative Common Neighbor Analysis (a-CAN), Centrosymmetry analysis and the dislocation analysis tool (DXA) as implemented in the OVITO software [37]. For more details, see refs [38, 39].

3 Results and Discussion

3.1 Hardening like Effect

The hardening-like effect is an elastic stiffening of the bcc matrix beyond plastic yielding, which inhibits the bcc to hcp phase transition up to extremely high pressures [38, 40]. To illustrate this effect, we consider a ramp compression of a [001]-oriented single-crystal iron sample with a maximum piston velocity of 800 m/s during 30 ps. Figs. 1, 2 and 3 shows the spatial distributions of the longitudinal pressure (Fig. 1(a)) and shear stress (Fig. 1(b)), along with selected atomic snapshots representing the evolving deformation structure. The corresponding maximum longitudinal pressure, P_z, is approximately 48 GPa, with an associated loading rate on the order of $4.9\text{x}10^9\text{s}^{-1}$.

As expected, the sample deforms elastically until a pressure of about 12 GPa (see the first step-like feature in the wave profiles in Fig. 1a and 2 at 20 ps and the corresponding snapshot in Fig. 3. Plastic flow then occurs through the nucleation of planar defects, which evolve into stacking faults identified as deformation twins (maroon-colored atoms) [38]. No dislocation slip is observed, likely due to the high purity of the sample [41]. Upon further compression, the previously formed micro-twins begin to recede around 18 GPa, leading to a second step in the P_z profiles, as seen in Fig. 1(a) at 20 ps, and inducing a new

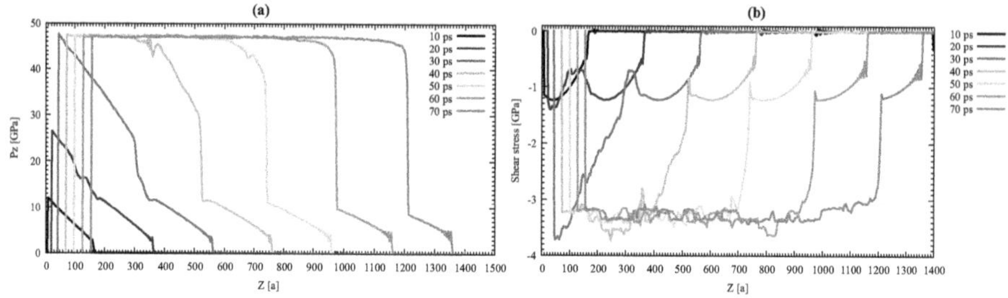

Fig. 1. Spatial distribution of (a) longitudinal pressure Pz and (b) shear stress in a sample ramp-compressed during 30 ps with a maximum piston velocity of 800 m/s.

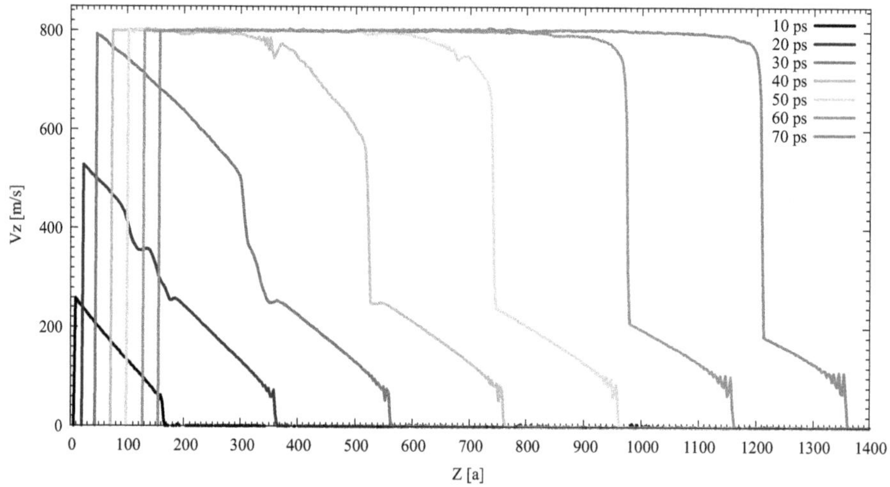

Fig. 2. Spatial distribution of the particle velocity in a sample ramp-compressed during 30 ps with a maximum piston velocity of 800 m/s

increase in the shear stress which had been relaxed by plastic deformation, see Fig. 2(b). A similar behavior was reported in other bcc materials, such as tantalum under shock compression at pressures exceeding 150 GPa [42]. The dynamic evolution of these twins was studied in [43], where it was shown that the longer the ramp compression time, the higher the twin fraction and the longer the duration of the subsequent receding phase.

Depending on the maximum pressure at the top of the ramp, the micro-twins can be either partially or completely receded, with a significant effect on both the dynamics of the bcc to hcp phase transition [38] and the resulting microstructure [44]. As mentioned above, this destruction of the micro-twins, with no dislocation activity detected in the sample, leads to an increase in the shear stress upon further compression, which is referred to as a hardening-like effect where the bcc matrix stiffens after plastic deformation.

This phenomenon has been analyzed in the context of classical theory of phase nucleation and growth in [38], where it was shown that in order to activate the structural bcc to hcp phase transition, one must either continue ramp compression to reach higher

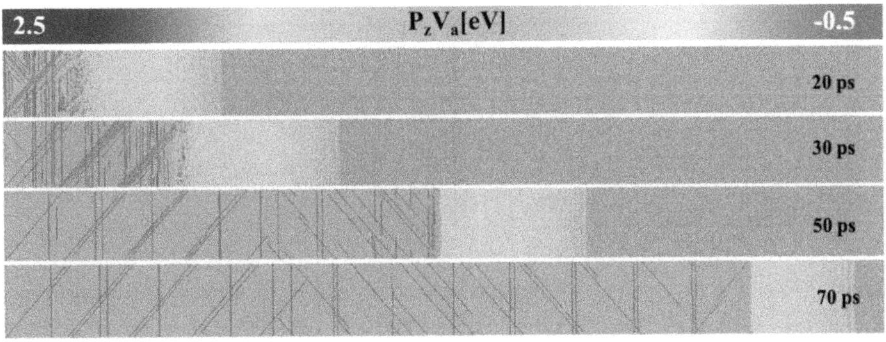

Fig. 3. Snapshot of the atomic configurations showing the structure of the deformation due to wave propagation: uncompressed BCC (orange), elastically deformed BCC (yellow), and plastically compressed BCC (green). Maroon-colored atoms represent stacking faults associated with plastic deformation, identified as micro-twins.

pressures of about 100 GPa or allow the ramp wave to steepen into a shock front. Indeed, because the sound speed increases with pressure, the ramp wave progressively steepens during its propagation [45–47], transforming the plastic deformation wave (P_1 wave) into an abrupt front approaching a shock wave. During this steepening process, for a sufficiently high maximum piston velocity, the bcc to hcp phase transition is triggered when the plastic deformation strain rate reaches a critical value that depends on the ramp rise time [43]. When this critical plastic strain rate is reached, the thermodynamic driving force is high enough for hcp embryos to grow and reach a critical size almost immediately behind the shock front [38, 39].

To illustrate these two scenarios for activating the bcc to hcp phase transition after the sample has undergone a hardening-like effect, the piston maximum velocity has been doubled (i.e., 1600 m/s), while keeping the ramp rise time at 30 ps. Figure 4 shows the spatial distributions of the longitudinal pressure and shear stress, with associated snapshots of atomic configurations presented in Fig. 5. The corresponding maximum pressure is about 140 GPa. Thus, here, two bcc to hcp phase transition fronts are observed [38, 43]: (i) at the top of the pressure ramp, in the region near the piston when the pressure reaches approximately 100 GPa, propagating from left to right; and (ii) behind the plastic wave P_1, where the ramp wave steepens into a shock front, in the middle of the sample, propagating in both directions. The second scenario is a consequence of the ramp wave steepening into a shock front. For details, see refs [38, 40, 43]

3.2 Release and Recompression Waves Emitted Front the BCC to HCP Phase Transition Front

For intermediate maximum piston velocities, only the second of the two scenarios above can be observed if the propagation distance is sufficient. Thus, the bcc to hcp phase transition will be activated in the middle of the sample, and it will lead to the emission of characteristic release and recompression waves from the bcc to hcp phase transition front. To illustrate this phenomenon, let us consider the same ramp piston maximum velocity of 800 m/s as in Figs. 1, 2 and 3. However, the ramp rise time is reduced by half (i.e.,

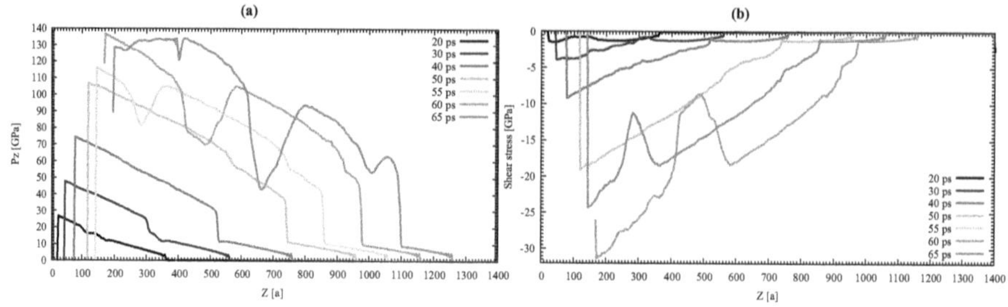

Fig. 4. Spatial distribution of (a) longitudinal pressure *Pz* and (b) shear stress in a sample ramp-compressed during 30 ps with a maximum piston velocity of 1600 m/s.

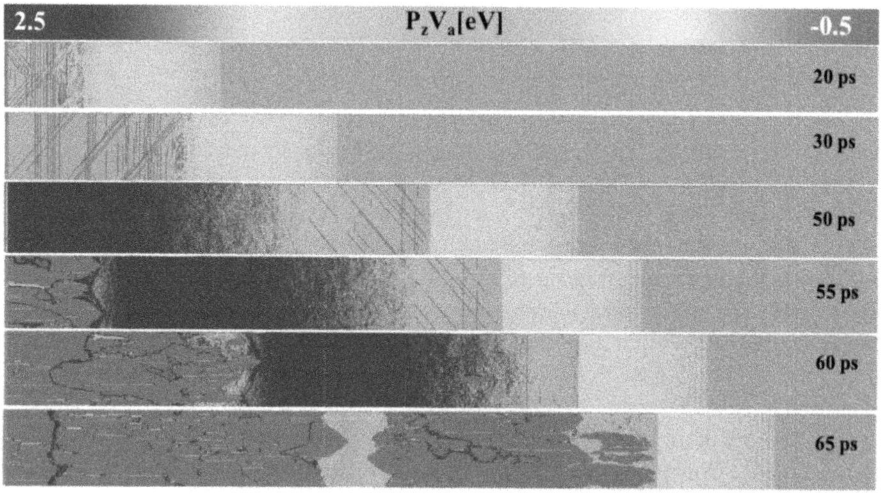

Fig. 5. Snapshot of the atomic configurations showing the structure of the deformation due to wave propagation: uncompressed BCC (orange), elastically deformed BCC (yellow), plastically compressed BCC (green), and the region subjected to a hardening-like effect (blue). Maroon-colored atoms represent stacking faults associated with plastic deformation, identified as micro-twins, while violet and cyan-colored atoms correspond to atoms in the HCP and FCC phases, respectively.

15 ps) to accelerate the ramp wave steepening during its propagation. Figure 6 shows the longitudinal pressure and shear stress at selected simulation times, while Figs. 7 and 8 show the corresponding profiles of the particle velocity and snapshots of the atomic configurations, respectively. The ramp maximum pressure is still 48 GPa, which is less than the micro-twins' complete receding threshold of about 60 GPa [38].

The associated maximum shear stress is about 3.2 GPa. Thus, at 20 ps, i.e., 5 ps after the ramp pressure reaches its maximum, micro-twins subsist in the region near the piston. At later times, the plastic wave (P_1) evolves into a shock front during its propagation, leading to a single sharp front behind the elastic precursor (see profile and snapshot at 50 ps in Figs. 6(a) and 7). Therefore, the high thermodynamic force associated with the shock wave activates the structural bcc to hcp phase transformation behind the P_1

wave. When the structural bcc to hcp phase transition starts, the shear stress is completely relaxed, even over-relaxed, behind the P_1 wave. Interestingly, release and recompression waves are emitted from the phase transition front (see Fig. 7) [39].

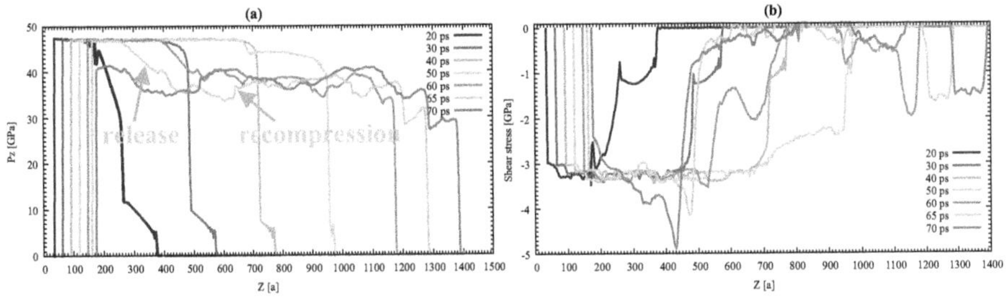

Fig. 6. Spatial distribution of (a) longitudinal pressure *Pz* and (b) shear stress in a sample ramp-compressed during 15 ps with a maximum piston velocity of 800 m/s. The pressure profile at 65 ps shows specific release and recompression waves emitted from the phase transition region, as fully detailed in Ref. [39].

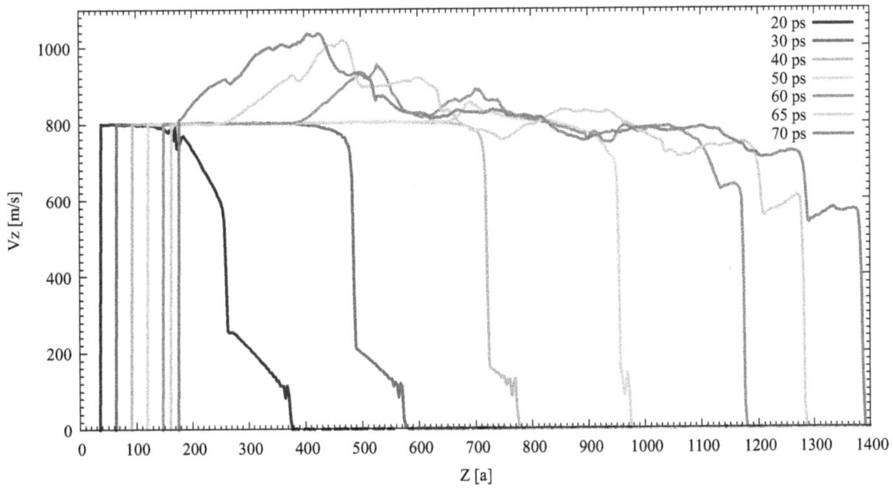

Fig. 7. Spatial distribution of the particle velocity *Vz* corresponding to the pressure and shear stress of Fig. 5.

Since the ramp compression required to reach a given pressure is greater than the shock compression [47, 48], the observed release wave might be attributed to a partial unloading from the isentrope to the Hugoniot state during shock formation. However, simulations at a lower piston maximum velocity of 600 m/s, where the maximum pressure was not enough to activate a phase transition, did not lead to the emission of release and recompression waves [44]. Consequently, the observation of these waves is due to the activation of the bcc to hcp phase transition in the middle of the sample. It can be

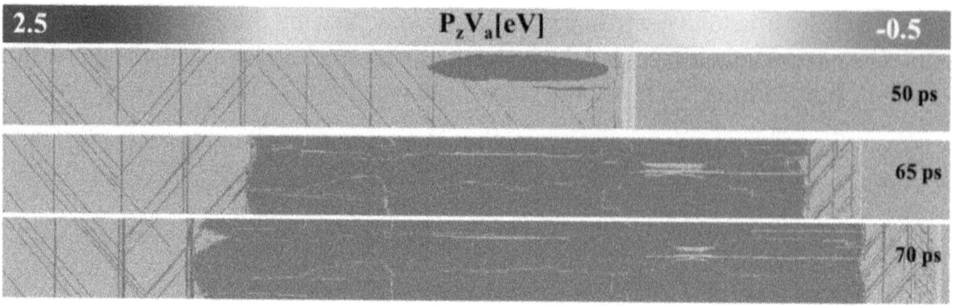

Fig. 8. Snapshot of the atomic configurations showing the activation of the bcc to hcp phase transition in the middle of sample.

interpreted through a phenomenological analysis based on the impedance mismatching method [45, 46]. Indeed, since the phase transition has a finite kinetics time, its activation in the middle of the sample leads to the formation of a mixed-phase zone, where bcc and hcp phases coexist [13, 16]. This mixed bcc and hcp phases zone separates the pure bcc region on the left side of the sample from the pure hcp region on the right, where the transformation has reached completion. It is well known that such a mixed-phase region generally exhibits lower density and reduced shock velocity compared to both the parent and product phases. In the pressure–particle velocity space, this is reflected by a flattening of the shock adiabat for the two-phase mixture, which lies below the Hugoniot curves of the pure phases [3, 49].

As a result, two interfaces are introduced within the sample, where shock impedance mismatches occur: the first between the pure bcc and the bcc/hcp mixed phase, and the second between the bcc/hcp mixed phase and the pure hcp phase. This scenario is analogous to shock propagation through a layered system with alternating hard and soft materials [45].

Figure 9 represents a schematic diagram of shock propagation through the sample, with the inset on the right illustrating the two interfaces. When an incident shock wave with a particle velocity u_p and a pressure P_{s1} reaches the first interface (pure bcc to bcc/hcp), since the mixed bcc/hcp phase has lower shock impedance than the pure bcc phase, it must unload isentropically to a lower pressure P_1 corresponding to a higher particle velocity u_1. This corresponds to a transmission of shock wave from harder to softer material. Thus, a reflected release wave is generated that propagates backward into the pure bcc region, while a shock wave with lower intensity (pressure) is transmitted into the mixed bcc/hcp region. This transmitted shock carries the mixed bcc/hcp phase zone from ambient conditions (zero pressure and zero particle velocity) to a point on the Hugoniot corresponding to the pressure P_1, which is given by the intersection of the release isentrope with the shock adiabat corresponding to the mixed state x_1.

Upon reaching the second interface (bcc/hcp mixed phase to pure hcp), the reverse situation occurs: transmission from softer to harder material. Thus, the transmitted shock continues into the pure hcp region, while a recompression wave is reflected back into the mixed-phase zone. This recompression wave carries the mixed bcc/hcp phase from the lower pressure $\mathrm{P_{s1}}$ to a higher pressure $\mathrm{P_{hcp}}$ corresponding to the particle velocity $\mathrm{u_p}$. Thus, upon the activation of bcc to hcp phase transformation, two waves, release

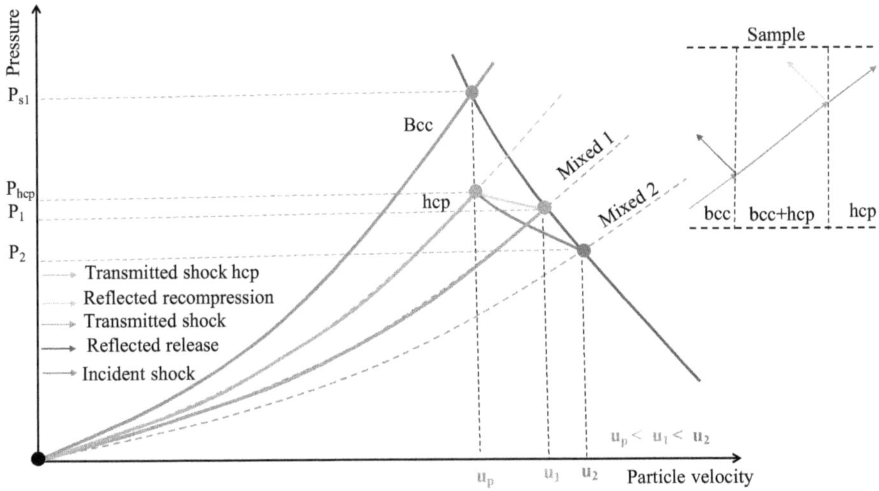

Fig. 9. Schematic diagram showing the waves' impedance mismatch through the sample. The inset on the right represents the virtual interfaces generated by the bcc to hcp phase transition activation.

and recompression waves, are generated from the phase transition front that propagates backward.

Because the mixed-phase region evolves over time due to the transformation kinetics, the amplitudes and timing of these waves are governed by the rate and extent of phase transformation. These kinetics dictate the thickness and phase proportions (x) within the mixed region. Thus, upon the evolving of the phase transition the release wave accelerates while the recompression steepens as illustrated in Fig. 8, see also Ref [39].

Figures 10 and 11 display isocontours of pressure P_Z and particle velocity in a space-time diagram, extracted from the simulation shown in Figs. 6, 7, and 8 (maximum piston velocity of 800 m/s with 15 ps ramp rise time). These diagrams visually represent the wave propagation and evolution within the sample over time. They show the initial elastic precursor followed by the plastic wave (P_1), which steepens into a shock front as it propagates. The diagrams also illustrate the activation of the bcc to hcp phase transition in the middle of the sample and the subsequent emission of a release and recompression wave from the phase transition front. This provides a clear visualization of the dynamic process, including the temporal and spatial relationships between the different wave fronts and material states. These results prove that atomistic simulations, in the accuracy limit of the interatomic potential, can capture fine features of the dynamic deformation of solid materials.

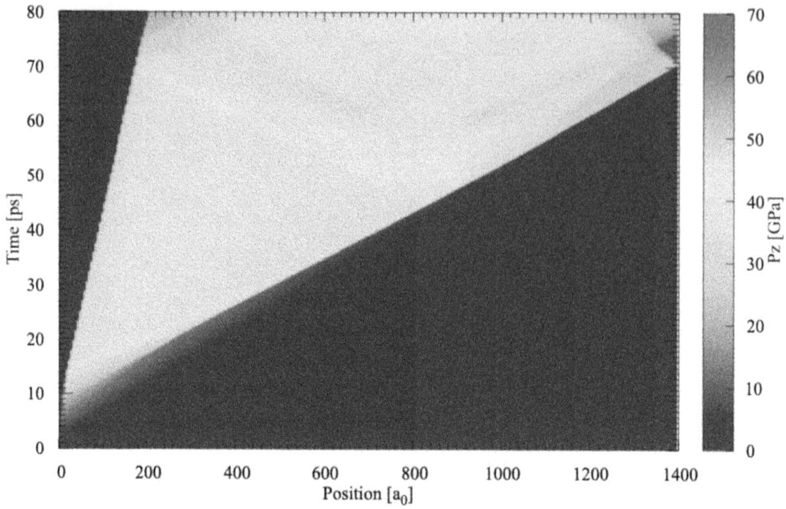

Fig. 10. Isocontours of pressure P_z in a space-time diagram extracted from the simulation shown in Figs. 5, 6 and 7 (maximum piston velocity of 800 m/s).

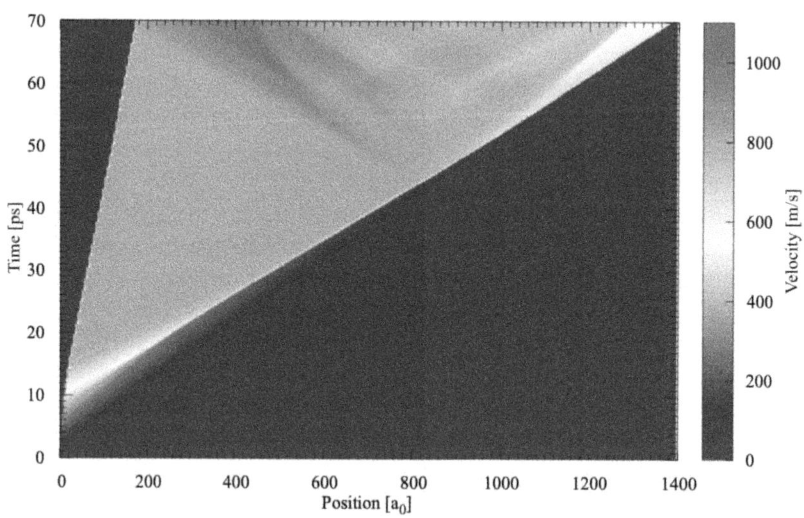

Fig. 11. Isocontours of particle velocity V_z in a space-time diagram extracted from the simulation shown in Figs. 5, 6 and 7 (maximum piston velocity of 800 m/s).

4 Conclusion

In conclusion, this proceeding summarizes our recent work on the dynamic response of single-crystal iron explored at the atomic level. Specifically, two new aspects were discussed (see Refs. 38 and 39 for more details):

(i) A hardening-like effect occurring upon ramp-compression beyond the elastic limit, that shifts the bcc to hcp phase transition to higher pressure. To activate this phase transition, one must either continue ramp compression to reach higher pressures of about 100 GPa or allow the ramp wave to steepen into a shock front.
(ii) Release and recompression waves emitted from the phase transition front. When the phase transition is activated in the middle of the sample due to ramp steepening into a shock wave, interfaces between parent, daughter, and mixed phase regions are introduced within the sample where shock impedance mismatch occurs. Thus, release and recompression waves are generated that propagate backward into the pure bcc region and the mixed bcc/hcp region, respectively.

These phenomena illustrate the capability of atomic-level simulations to capture not only the details of wave profiles and deformation mechanisms but also new fine features yet to be uncovered experimentally.

Acknowledgments. "This work was carried out with the aid of a grant from UNESCO-TWAS financed by the Swedish International Development Cooperation Agency (Sida). The views expressed herein do not necessarily represent those of UNESCO-TWAS, Sida or its Board of Governors"

This work was also partially funded by the French program CNRS-Africa.

Computations were performed on the supercomputer facilities of the Mésocentre de calcul de Poitou Charentes (France).

References

1. Bancroft, D., Peterson, E.L., Minshall, S.: Polymorphism of iron at high pressure. J. Appl. Phys. **27**, 291–298 (1956)
2. Barker, L.M., Hollenbach, R.E.: Shock wave study of the a-e transition in iron. J. Appl. Phys. **45**, 4872–4887 (1974)
3. Duvall, G.E., Graham, R.A.: Phase transitions under shock-wave loading. Rev. Mod. Phys. **49**(3), 523–581 (1977)
4. Boettger, J.C., Wallace, D.C.: Metastability and dynamics of the shock-induced phase transition in iron. Phys. Rev. B. **55**, 2840–2849 (1997)
5. Kalantar, D.H., et al.: Direct observation of the alpha-epsilon transition in shock-compressed iron via nanosecond X-ray diffraction. Phys. Rev. Lett. **95**, 075502 (2005)
6. Hawreliak, J., et al.: Analysis of the X-ray diffraction signal for the α–ε transition in shock-compressed iron: simulation and experiment. Phys. Rev. B. **74**, 184107 (2006)
7. Hawreliak, J.A., Kalantar, D.H., Stölken, J.S., Remington, B.A., Lorenzana, H.E., Wark, J.S.: High-pressure nanocrystalline structure of a shock-compressed single crystal of iron. Phys. Rev. B. **78**, 220101(R) (2008)
8. Lorenzana, H.E., et al.: Shocked materials at the intersection of experiment and simulation. Sci. Model. Simul. **15**, 159–186 (2008)

9. Bastea, M., Bastea, S., Becker, R.: High pressure phase transformation in iron under fast compression. Appl. Phys. Lett. **95**, 241911 (2009)
10. Jensen, B.J., et al.: Dynamic compression of iron single crystals. AIP Conf. Proc. **845**, 232–235 (2006)
11. Jensen, B.J., Gray III, G.T., Hixson, R.S. : Direct measurements of the a–e transition stress and kinetics for shocked iron. J. Appl. Phys. **105**, 103502 (2009)
12. Smith, R.F., et al.: High strain-rate plastic flow in Al and Fe. J. Appl. Phys. **110**, 123515 (2011)
13. Smith, R.F., et al.: Time-dependence of the alpha to epsilon phase transformation in iron. J. Appl. Phys. **114**, 223507 (2013)
14. Crowhurst, J.C., et al.: The phase transition in iron at strain rates up to $10^9 s^{-1}$. J. Appl. Phys. **115**, 113506 (2014)
15. Zaretsky, E.A., Kanel, G.I.: Yield stress, polymorphic transformation, and spall fracture of shock-loaded iron in various structural states and at various temperatures. J. Appl. Phys. **117**, 195901 (2015)
16. Amadou, N., et al.: Kinetics of the iron α-ε phase transition at high-strain rates: experiment and model. Phys. Rev. B. **93**, 214108 (2016)
17. Amadou, N., et al.: Laser-driven ramp compression to investigate and model dynamic response of iron at high strain rates. Meta **6**, 320 (2016)
18. Yaakobi, B., et al.: EXAFS measurement of iron bcc-to-hcp phase transformation in nanosecond-laser shocks. Phys. Rev. Lett. **95**, 075501 (2005)
19. Hwang, J., et al.: Subnanosecond phase transition dynamics in laser-shocked iron. Sci. Adv. **6**, eaaz5132 (2020)
20. Johnson, J.N., Rohde, R.W.: Dynamic deformation twinning in shock-loaded iron. J. Appl. Phys. **42**(11), 4171–4180 (1971)
21. de Rességuier, T., Hallouin, M.: Stress relaxation and precursor decay in laser shock-loaded iron. J. Appl. Phys. **84**, 1932–1938 (1998)
22. Barton, N.R., et al.: A multiscale strength model for extreme loading conditions. J. Appl. Phys. **109**, 073501 (2011)
23. Hayes, D.B.: Wave propagation in a condensed medium with N transforming phases: application to solid-I–solid-II–liquid bismuth. J. Appl. Phys. **46**, 3438 (1975)
24. de Rességuier, T., Hallouin, M.: Interaction of two laser shocks inside iron samples. J. Appl. Phys. **90**(8), 4377–4384 (2001)
25. Avrami, M.: Kinetics of phase change. I general theory. J. Chem. Phys. **7**, 1103 (1939)
26. Avrami, M.: Kinetics of phase change. II. Transformation-time relations for random distribution of nuclei. J. Chem. Phys. **8**, 212–224 (1940)
27. Barton, N.R., Benson, D.J., Becker, R.: Crystal level continuum modelling of phase transformations: the α ↔ ε transformation in iron. Model. Simul. Mater. Sci. Eng. **13**(5), 707–731 (2005)
28. Kadau, K., Germann, T.C., Lomdahl, P.S., Holian, B.L.: Atomistic simulations of shock-induced transformations and their orientation dependence in bcc Fe single crystals. Phys. Rev. B. **72**, 064120 (2005)
29. Kadau, K., et al.: Shock waves in polycrystalline iron. Phys. Rev. Lett. **98**, 135701 (2007)
30. Gunkelmann, N., et al.: Shock waves in polycrystalline iron: plasticity and phase transitions. Phys. Rev. B. **89**, 140102(R) (2014)
31. Thompson, A.P., et al.: LAMMPS – a flexible simulation tool for particle-based materials modeling at the atomic, meso, and continuum scales. Comput. Phys. Commun. **271**, 108171 (2022)
32. Daw, M.S., Baskes, M.I.: Semiempirical, quantum mechanical calculation of hydrogen embrittlement in metals. Phys. Rev. Lett. **50**, 1285–1288 (1983)

33. Foiles, S.M., Baskes, M.I., Daw, M.S.: Embedded-atom-method functions for the fcc metals cu, ag, au, Ni, Pd, Pt, and their alloys. Phys. Rev. B. **33**, 7983–7991 (1986)
34. Gunkelmann, N., Bringa, E.M., Kang, K., Ackland, G.J., Ruestes, C.J., Urbassek, H.M.: Polycrystalline iron under compression: plasticity and phase transitions. Phys. Rev. B. **86**, 144111 (2012)
35. Gunkelmann, N., Tramontina, D.R., Bringa, E.M., Urbassek, H.M.: Interplay of plasticity and phase transformation in shock wave propagation in nanocrystalline iron. New J. Phys. **16**, 093032 (2014)
36. Luu, H.-T., Gunkelmann, N.: Pressure-induced phase transformations in Fe-C: molecular dynamics approach. Comput. Mater. Sci. **162**, 295–303 (2019)
37. Stukowski, A.: Visualization and analysis of atomistic simulation data with OVITO–the open visualization tool. Model. Simul. Mater. Sci. Eng. **18**, 015012 (2010)
38. Amadou, N., de Resseguier, T., Dragon, A., Brambrink, E.: Coupling between plasticity and phase transition in shock- and ramp-compressed single-crystal iron. Phys. Rev. B. **98**, 024104 (2018)
39. Amadou, N., de Rességuier, T.: Pressure waves induced by the bcc-hcp phase transition in dynamically loaded single crystal iron. Comput. Mater. Sci. **247**, 113559 (2025)
40. Amadou, N., de Rességuier, T., Dragon, A.: Coupling between plasticity and phase transition in single crystal iron at ultra-high strain rate. AIP Conf. Proc. **2272**, 070001 (2020)
41. Amadou, N., de Rességuier, T., Dragon, A., Brambrink, E.: Effects of orientation, lattice defects and temperature on plasticity and phase transition in ramp-compressed single crystal iron. Comput. Mater. Sci. **172**, 109318 (2020)
42. Wehrenberg, C.E., et al.: In situ X-ray diffraction measurement of shock-wave-driven twinning and lattice dynamics. Nature **550**, 496–500 (2017)
43. Amadou, N., Ayouba Abdoulaye, A.R.A., De Rességuier, T., Dragon, A.: Strain-rate dependence of plasticity and phase transition in [001]-oriented single-crystal iron. Crystals **13**, 250 (2023)
44. Nassirou Hassan, M.A., Amadou, N., Moussa Hassane, A., Batouré, A., Adamou, I., de Rességuier, T.: Kinetics of the bcc to hcp phase transition in [001]-oriented single crystal iron under ramp compression. Computat. Condens. Matter **41**, e00965 (2024)
45. Meyers, M.A.: Dynamic Behavior of Materials. John Wiley & Sons, Hoboken, NJ, USA (1994)
46. Zel'Dovich, Y.B., Raizer, Y.P.: Physics of Shock Waves and High-Temperature Hydrodynamic Phenomena, Dover (2002)
47. Swift, D.C., Kraus, R.G., Loomis, E.N., Hicks, D.G., McNaney, J.M., Johnson, R.P.: Shock formation and the ideal shape of ramp compression waves. Phys. Rev. E. **78**(6), 066115 (2008)
48. Hayes, D.B., Hall, C.A., Asay, J.R., Knudson, M.D.: Measurement of the compression isentrope for 6061-T6 aluminum to 185 GPa and 46. J. Appl. Phys. **96**(10), 5520–5527 (2004)
49. Kanel, G.I., Razorenov, S.V., Fortov, V.E.: Polymorphic transformations and phase transitions in shock-compressed solids. In: Shock-Wave Phenomena and the Properties of Condensed Matter. Springer, New York (2004)

A Parametric Design Approach to Spider Web-Inspired 3D Printed Load-Bearing Structures

Gaurab Sundar Dutta[1](✉), Krishnachandran Chandrasekharan Nair Jayalekshmi[1], and Uwe Wolfram[2]

[1] Institute of Polymer Materials and Plastic Engineering, Clausthal University of Technology, Clausthal-Zellerfeld 38678, Germany
{gaurab.sundar.dutta, krishnachandran.chandrasekharan.nair.jayalekshmi}@tu-clausthal.de

[2] The Institute of Materials Science and Engineering, Clausthal-Zellerfeld 38678, Germany
uwe.wolfram@tu-clausthal.de

Abstract. Spider webs are captivating natural structures that have fascinated researchers for their remarkable properties. The silk produced by spiders has been well-documented for its exceptionally lightweight yet strong characteristics, while the intricate 2D and 3D web designs are notable for their span over large areas with efficient and minimal material usage while maintaining optimal structural integrity. Spiders achieve this type of structure by strategically distributing material resources, thus making the structure naturally optimized for a given condition, leaving a compelling challenge for researchers to recreate such a form for different use cases following biomimicry.

This work focuses on 2D orb-web designs, which are spiral formations connected via radial members. The idea is to develop a parametric model that mimics and explores such a shape, and the model can then be enhanced to include other design criteria. Initial case studies will focus on 3D printed web structures and their behavior under compressive loading conditions. For simulation, a particle-spring-based solver integrated with an evolutionary programming tool will be employed, generating deformation energy as output. Once a suitable correlation between experiment and simulation is met, the work will be extended to use cases, considering mapping such shapes onto other curved geometric forms.

A direct application of this research will be to utilize such structures as biomedical inserts. The idea will be explored in the last phase for a skull insert. Ultimately, this work aims to understand the underlying principles of spider-web structure and develop a bottom-up design strategy to replicate these natural forms.

Keywords: Spider Webs · Biomimicry · Parametric Modeling · Evolutionary Programming · 3D Printing · Biomedical Insert

N. Merkert and A. Herzog (Eds.): SimScience 2025, CCIS 3027, pp. 15–26, 2026.
https://doi.org/10.1007/978-3-032-29720-4_2

1 Introduction

In the intricate and complex domain of nature, numerous examples of ingenious designs exist that have evolved over millions of years to solve complex problems and adapt to changing environments. From the majestic architecture of termite mounds to the remarkable camouflage abilities of chameleons and the progressive growth of plants towards light sources, nature has consistently demonstrated an unparalleled capacity for innovation and efficiency. The branch of science and technology dedicated to investigating such instances in nature to create novel solutions for real-life problems is called biomimicry. This design philosophy leverages nature's efficient and adaptive strategies to solve human challenges in sustainable and elegant ways [1, 2].

One such example is the study of spiderwebs, which, with their delicate patterns and networks, have fascinated scientists and engineers for centuries. In nature, spiders use their webs made of spider-silk [3] for various purposes, including web-building, wrapping prey, protecting offspring, and as a lifeline to ensure a safe escape from predators. Consequently, there exist different types of webs, with more than 130 identified so far, made out of seven different types of silks, each having its unique characteristics [4–6].

The orb web is one such example, which excels in exhibiting a high degree of geometrical regularity, allowing for structural stability even at a larger scale. This makes the orb web particularly interesting for engineering designs that aim to maximize strength and flexibility while using minimum material. The orb web is typically composed of flexible spiral threads supported and held together by radial lines joining the center of the web to the anchor points. The assembly is then supported by an outer frame, which is usually stiffer compared to the other parts of the assembly [7, 8] (Fig. 1).

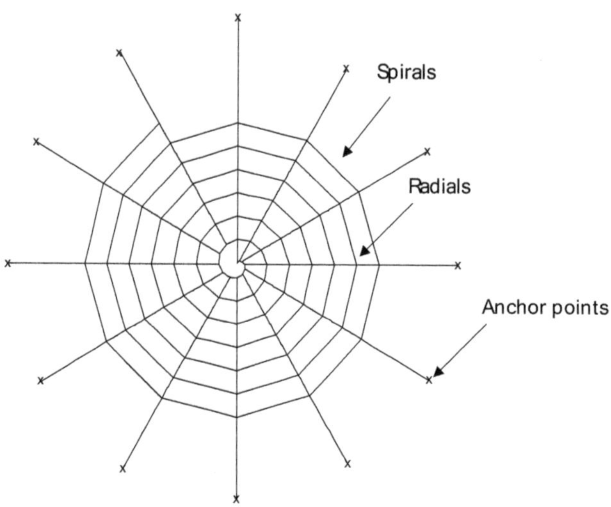

Fig. 1. Schematic diagram of an orb-web construction.

Depending on the arrangement of the spirals, such a system becomes highly efficient in energy absorption and dissipation, allowing the web to be flexible under a high number of loads. Several engineering studies have been conducted in the past few decades to understand and exploit the behavior of such arrangements [9–11].

This work looks into the feasibility of generating such a pattern using Fused Deposition Modeling (FDM) printing, delving into the generative design of such a geometry for specific engineering applications.

2 Methodology

2.1 Case Study Definitions

To understand and exploit the architectural and mechanical logic embedded in orb web geometry through the lens of computational design and mechanical testing, the work was divided into different phases:

Phase 1: Study of 2D orb web

This phase involved the parametric modelling and simulation of spider web geometries to analyze their mechanical behavior under various loading conditions. Using finite element analysis (FEA) and 3D printing, the effects of variations in web patterns on stress distribution, deformation, and failure points were observed. The aim was to extract performance principles that can inform further complex applications.

Phase 2: Application to dome structures

Building on the insights obtained from phase 1, this phase integrates orb web patterns mapped into three-dimensional dome structures, contrasting them with the traditional hexagonal dome configurations. Through simulations and physical prototyping, the impact of the geometrical configuration of orb web structure on the structural integrity, load efficiency, and aesthetic potential of a double curvature architectural form was assessed.

Phase 3: Outlook into the feasible application in the biomedical field

Once the proof of concept for a web-based structure was established, the know-how was further extended into biomedical applications, e.g., bone inserts. For this purpose, an in-house 3D scan was performed to generate the model of a human head, and in a computational model, a section was replaced with a mapped web pattern, which was further optimized for applied load.

2.2 Design and Fabrication

The first step of this study was the parametric design of the 2D orb web. Rhino [12] modelling tool with the Grasshopper [13] visual programming domain was used for this purpose. For phase 1, a secondary frame, a typical hexagonal geometry was always kept constant for all the combinations of radials. The length of the spirals was also kept fixed, and the final mass of each sample was kept constant by changing the thickness of each member.

Four such designs were generated to observe the effect of radial members, with every other design parameter being fixed. The designs were then exported to a standard FDM printer to produce prototypes with PLA material for mechanical loads. Details of each type of sample are mentioned in the following Table 1, and the models are illustrated in Fig. 2.

Table 1. Different types of samples for testing in phase 1.

	A	B	C	D
No. of radials	6	12	18	24
Length of spiral	88 cm	88 cm	88 cm	88 cm
Thickness of members	1.4 mm	1.2 mm	1.0 mm	0.8 mm
Secondary frame thickness	2 mm	2 mm	2 mm	2 mm
Overall mass	8 g	8 g	8 g	8 g

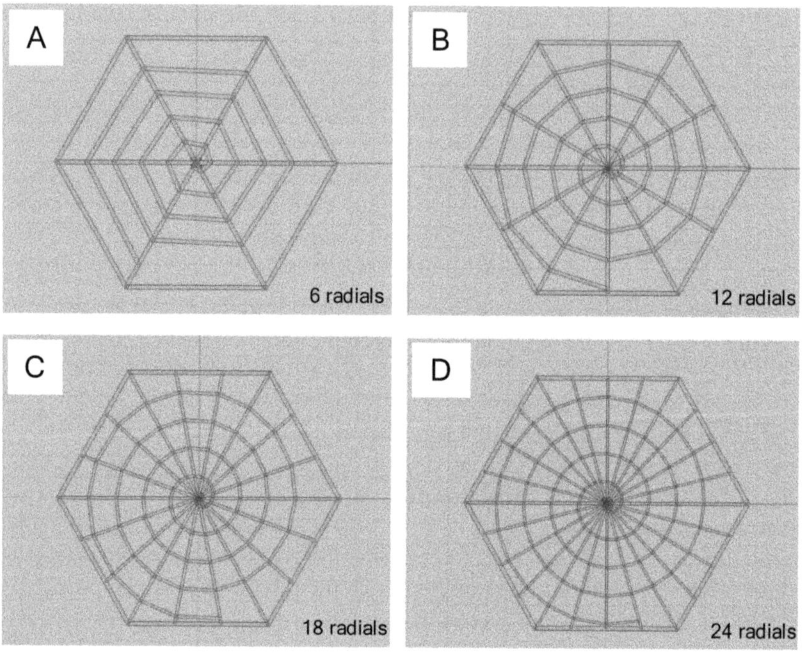

Fig. 2. Generated web geometries according to the specifications mentioned in Table 1.

For phase 2, the primary goal was to find a case study to map the web geometry on a curved surface to observe the geometrical behavior of the shape under compression load. For this purpose, a basic dome shape, as shown in Fig. 3, was constructed in Rhino+Grasshopper domain, and web geometries were mapped onto it.

The following constraints were implemented to generate the orb web structures:

- Fixed base: The base of the dome remained unchanged across all samples.
- Constant thickness: Both radial and spiral threads were maintained at consistent thickness across all domes.
- Fixed total mass: The final mass of each sample was kept constant by varying the number of radials and spirals.

Fig. 3. Dome surface to map.

Based on the results obtained from the analysis as described in Sect. 3.2, three web geometries were generated for printing. Finally, to have a qualitative comparison, a hexagonal pattern was mapped on the dome surface, and prototypes were generated using 3D printing with PLA material. Details of the prototypes are shown in Fig. 4.

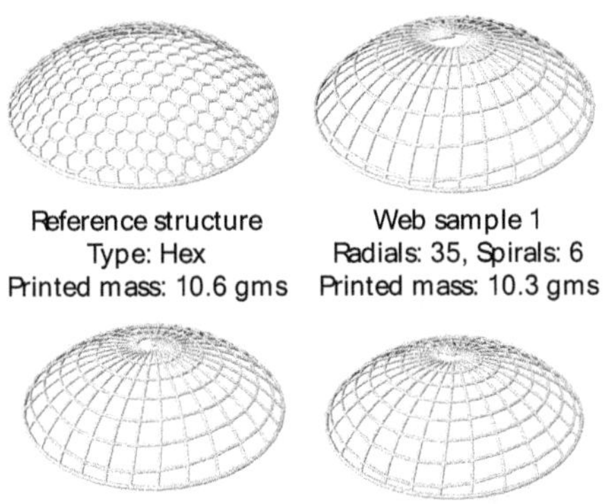

Fig. 4. Generated models of mapped surfaces as test prototypes.

3 Results and Analysis

3.1 Phase 1

In this phase, the surface strength of the 2D printed prototypes was measured using a universal testing machine (UTM) capable of applying compressive axial load. The setup was modified by elevating and placing the printed samples on a hollow cylinder made of stainless steel, and the load was applied at the center, with boundaries of the samples resting on the edges of the cylinder, as can be seen in Fig. 5.

Tests were performed under 3 different feed rates, and the responses were noted for each type of prototype until either the samples broke or slippage in the boundaries occurred. This controlled variation in feed rates provided insight into the mechanical behavior of the design under different loading conditions, allowing for a comprehensive

Fig. 5. Axial loading set-up using a UTM.

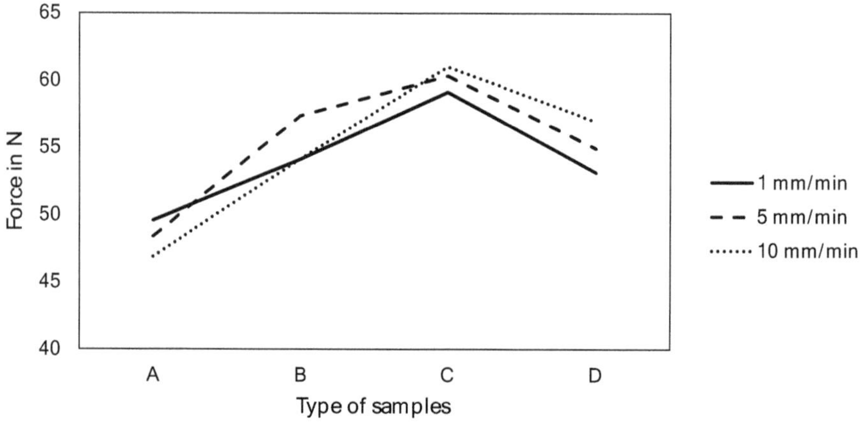

Fig. 6. Axial loading test results under different feed rates.

analysis of the sample's performance. Figure 6 illustrates the comparison between the force response for each type of sample under different feed rates.

The obtained results suggest that the strength across the geometries and feed rates increase as the number of radials is increased until model C. Beyond this point, the strength drastically decreases, resulting in a weaker prototype. This is probably due to the low thickness (0.8 mm) of the members in model D, resulting in the failure along the joints in the 3D printed samples.

The results were further validated through static structural FEA in ANSYS Workbench. A fixed support to the secondary frame and a 100 mm displacement load at the center of each sample were applied. The results, as described in Fig. 7, show that the outcome of the simulation is closely aligned with the experimental data, exhibiting the lowest von Mises stress for model C.

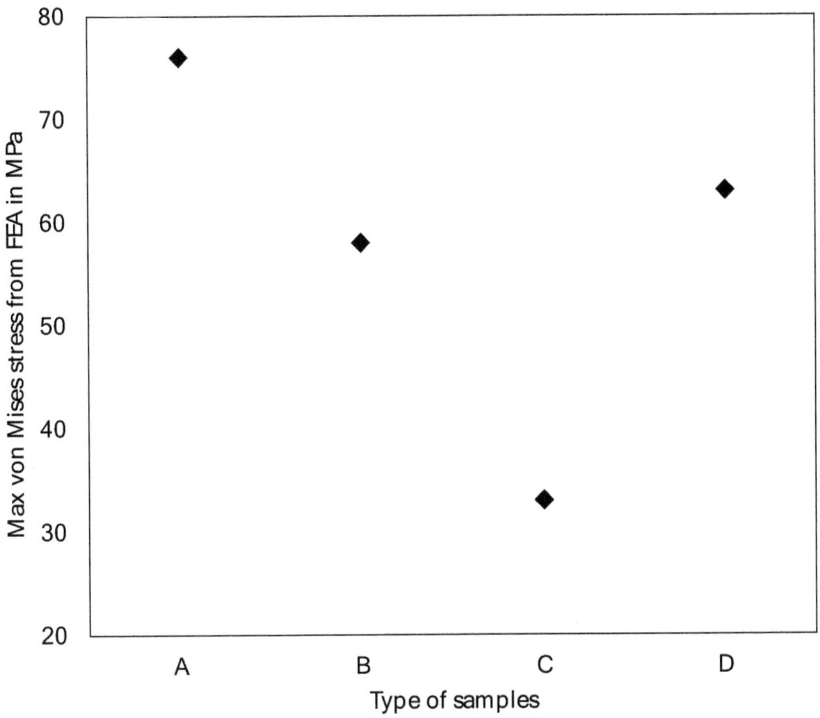

Fig. 7. FEA results obtained from Ansys Workbench for displacement loading condition.

3.2 Phase 2

Once a basic understanding of the different parameters of the orb web was achieved, the parametric model was extended to generate dome-shaped samples. Before mapping as described in Sect. 2.2, the geometries generated via the parametric model were subjected to Particle-spring analysis. The Particle-spring method is a standard deformation analysis tool, where the geometry is divided into a collection of lumped masses called particles, connected by linear elastic springs. The stiffness of the springs allows the shape to deform under applied load, and the energy in the deformed system is calculated. These values can further be given as input to an evolutionary solver for a qualitative comparison between different geometries. Schematics of the method are illustrated in Fig. 8. Within the Grasshopper domain, the Kangaroo Physics [14] toolbox was utilized for deformation analysis, while Galapagos [15] served as the optimization tool.

Consequently, prototypes were generated as described in Sect. 2.2 and were subjected to compressive load with a fixed base. Figure 9 summarizes the experimental set-up, mechanical test and simulation results.

From the obtained results, it can be safely concluded that the web samples are much more stable as compared to the hex sample. Within the tested web samples, both sample 1 and sample 2 show a similar pattern of deformation, with sample 3 closely following. Thus, for a given system, a number of probable solutions can be proposed by this method, promoting generative designs.

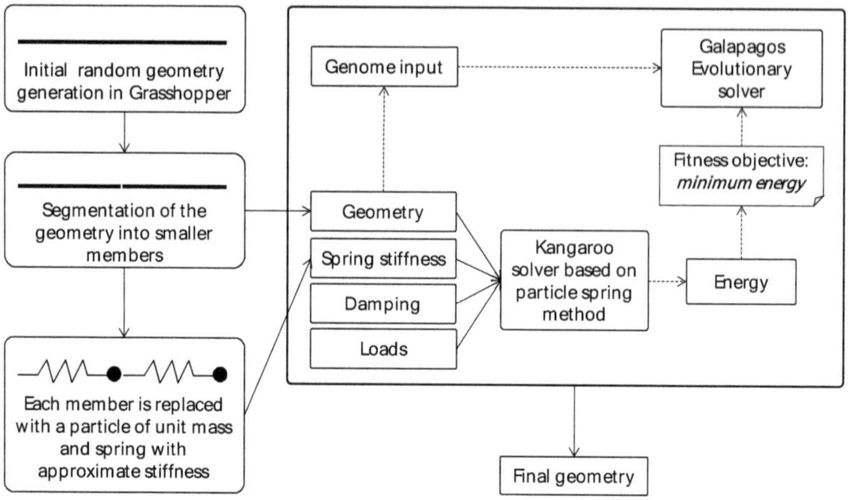

Fig. 8. Illustration of Particle spring analysis in Grasshopper.

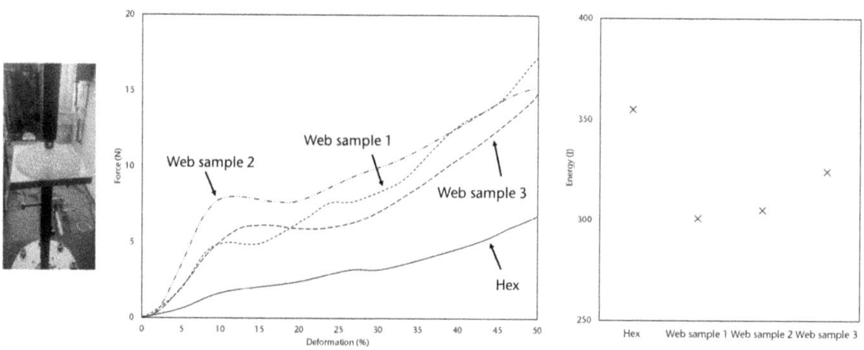

Fig. 9. (left) Representative image of the experimental setup; (middle) compression test results; (right) Particle-spring analysis results showing energy in the deformed systems.

3.3 Phase 3

As an outlook to the present work, the authors extended the idea into designing biomedical inserts. For design simplicity, a shape similar to a dome structure, as described in the previous section, was chosen, which turned out to be the frontal segment of the human skull. To have a preliminary estimation, an in-house 3D scan of the head, as shown in Fig. 10, was conducted, and the surface mesh was refined to get a smooth geometrical model imported into Rhino+Grasshopper domain.

Since material consumption would be a critical factor in such a use case, Octopus [16], a multi-objective evolutionary solver, was integrated into the generative model. Consequently, energy in the deformed configuration and the total length of the generated geometries became two key objectives to be considered. Consequently, a Pareto front was achieved, and two random solutions, as shown in Fig. 11, were chosen for

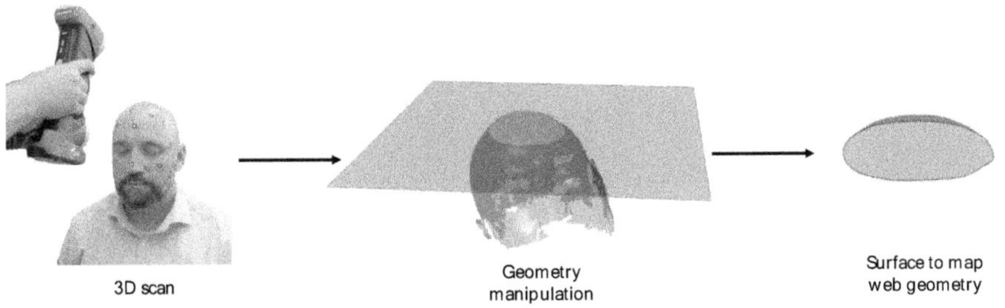

Fig. 10. Surface extraction workflow for biomedical application.

mapping. Before mapping, the Karamba 3D [17] finite element solver was used to validate the results obtained from the multi-objective optimizer, and respective deformation coefficients were calculated.

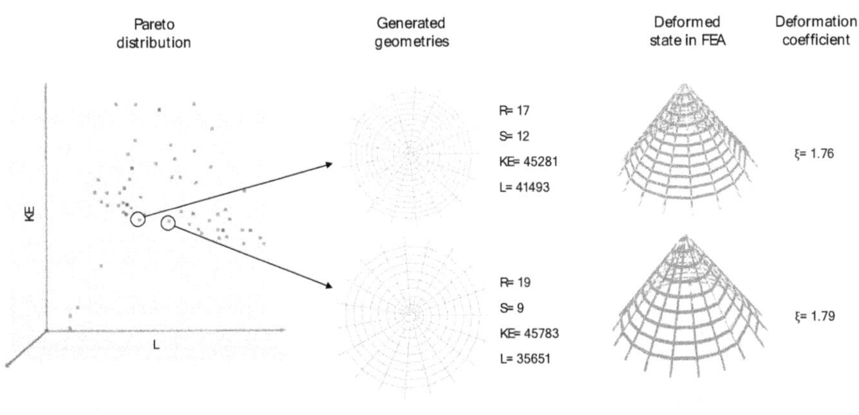

Fig. 11. (left) Pareto distribution obtained from the multi-objective solver; (middle) Two randomly selected geometries from the Pareto front and their radial, spiral, KE and total length specifications; (right) Surface FE results from Karamba 3D and deformation coefficient values.

The closer deformation coefficient [18] values suggest that both geometries can perform as a potential solution. Once finalized, the geometries were mapped back to the head cut section, as shown in Fig. 12, to gain a preliminary idea.

In the future, the work will be further extended to prepare prototypes from these generated models using standard medical-grade material via additive manufacturing [19–21]. Once structural tests are conducted and suitable results are obtained, the feasibility check will be done to use them as actual inserts.

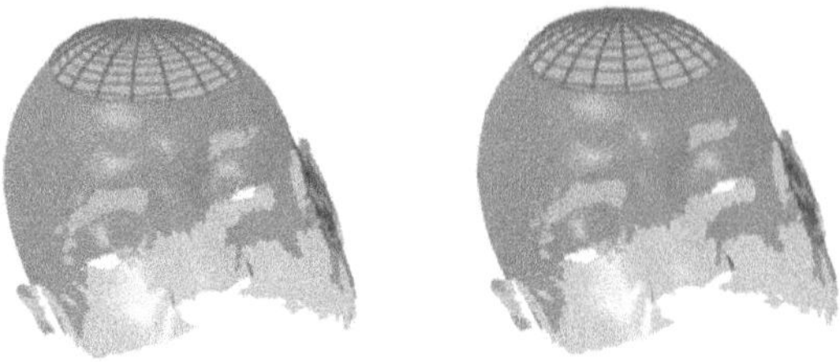

R=17, S= 12 R= 19, S= 9

Fig. 12. Final mapped geometries integrated with the head scan model.

4 Conclusion

The present work demonstrates the potential of orb web-inspired geometries in engineering and biomedical fields, showcasing their ability to provide enhanced stability and structural integrity. By combining parametric tools, additive manufacturing, and biomechanical inspiration, this research proposes a new framework for generative designs of load-efficient web-inspired designs.

A three-phase approach was chosen, where in the first phase, the parametric modeling and finite element simulation of spider web-inspired geometries were performed to analyze their mechanical behavior under axial compressive loading conditions. The results showed that the strength of the web increased with the number of radials until a certain point, beyond which it decreased. Since mass was kept constant, this resulted in low thickness of the members in the 3D printed samples, causing potential failure.

In the second phase, the orb web patterns were mapped onto a curved surface to observe the geometrical behavior of the shape under compression load. The experimental results showed that the web samples were more stable than the hexagonal sample. Moreover, the results also showed that multiple web-inspired structures can be found as generative solutions for the same boundary and loading conditions, resulting in similar structural responses. The results were further validated with Particle-spring analysis and evolutionary programming, observing the energy stored in each of the deformed systems.

In the last phase, the idea was extended to design biomedical inserts. The frontal segment of the human skull was chosen as a potential case study by conducting an in-house 3D scan of the head and by creating a geometrical model out of it. A multi-objective evolutionary solver was integrated into the generative model to consider material consumption as a critical factor. A Pareto front was achieved, and two random solutions were chosen for mapping. The surface-based finite element solver was used to validate the results obtained from the multi-objective optimizer. In the end, intricate web structures were generated, thus making advancements towards producing biomedical inserts.

For future works, the proposed methodology needs to be tested for different materials involving different printing technologies. Moreover, limitations in mesh-based slicing for G-code generation ignore the effect of directionality while printing material. Thus, a curve-based slicing technique needs to be adopted to realize the full potential of such structures, exploiting material anisotropy.

In the end, this work was limited only to orb-web geometry, and such methodology can be extended to find potential applications for other spider-web-inspired geometries, paving novel ways of design and manufacturing of smart materials inspired by biomimicry.

Acknowledgements. The authors would like to thank Leif Steuernagel and Fabian Hartkopf for their valuable inputs.

References

1. Benyus, J.M.: Biomimicry: innovation inspired by nature. Perennial (2002)
2. Bejan, A., Zane, J.P.: Design in Nature: how the Constructal Law Governs Evolution in Biology, Physics, Technology, and Social Organization. Doubleday (2012)
3. Römer, L., Scheibel, T.: The elaborate structure of spider silk: structure and function of a natural high-performance fiber. Prion. **2**(4), 154–161 (2008)
4. Blamires, S.J., Zhang, S., Tso, I.M.: Webs: diversity, structure and function. In: Behaviour and Ecology of Spiders: Contributions from the Neotropical Region, pp. 137–164. Springer International Publishing, Cham (2017)
5. Vollrath, F.: Spider silk: thousands of nano-filaments and dollops of sticky glue. Curr. Biol. **16**(21), R925–R927 (2006)
6. Blackledge, T.A., et al.: Reconstructing web evolution and spider diversification in the molecular era. Proc. Natl. Acad. Sci. **106**(13), 5229–5234 (2009)
7. Qin, Z., Compton, B.G., Lewis, J.A., Buehler, M.J.: Structural optimization of 3D-printed synthetic spider webs for high strength. Nat. Commun. **6**(1), 7038 (2015)
8. Soler, A., Zaera, R.: The secondary frame in spider orb webs: the detail that makes the difference. Sci. Rep. **6**(1), 31265 (2016)
9. Mortimer, B., Soler, A., Siviour, C.R., Zaera, R., Vollrath, F.: Tuning the instrument: sonic properties in the spider's web. J. R. Soc. Interface. **13**(122), 20160341 (2016)
10. Regassa, Y., Lemu, H.G., Sirabizuh, B., Rahimeto, S.: Studies on the geometrical design of spider webs for reinforced composite structures. J. Compos. Sci. **5**(2), 57 (2021)
11. Naderinejad, M., Junge, K., Hughes, J.: Exploration of the design of spiderweb-inspired structures for vibration-driven sensing. Biomimetics **8**(1), 111 (2023)
12. Rhinoceros 3D: https://www.rhino3d.com/. Accessed 25 Aug 2025
13. Grasshopper 3D: https://www.grasshopper3d.com/. Accessed 25 Aug 2025
14. Kangaroo Physics: https://www.food4rhino.com/app/kangaroo-physics. Accessed 25 Aug 2025
15. Galapagos Evolutionary Computing: https://www.grasshopper3d.com/group/galapagos. Accessed 25 Aug 2025
16. Octopus: https://www.food4rhino.com/en/app/octopus. Accessed 25 Aug 2025
17. Karamba 3D: https://www.karamba3d.com/. Accessed 25 Aug 2025
18. Dutta, G.S., Meiners, D., Merkert, N.: A study of free-form shape rationalization using biomimicry as inspiration. Polymers **15**(11), 2466 (2023)
19. Erbereli, R., de Camargo, I.L., Morais, M.M., Fortulan, C.A.: 3D printing of trabecular bone-mimetic structures by vat photopolymerization of bovine hydroxyapatite as a potential candidate for scaffolds. J. Braz. Soc. Mech. Sci. Eng. **44**(5), 170 (2022)
20. Morais, M.M., et al.: Fabrication and petrophysical characterization of artificial carbonate rocks with multiscale porosity sintered in a CO2 atmosphere. Geoenergy Sci. Eng. **229**, 212096 (2023)

21. Morais, M.M., de Camargo, I.L., Colombo, P., Fortulan, C.A.: Additive manufacturing of calcium carbonate parts through vat-photopolymerization and sintering in carbon dioxide atmosphere. Open Ceram. **14**, 100348 (2023)

The Digital Twin of Quality Control Labs of the Business Unit Nutrition & Health at BASF Ludwigshafen

Niklas Loges[1](✉) and Alexander Herzog[2]

[1] BASF SE Ludwigshafen, Nutrition & Health – Head of Quality Control, Carl-Bosch-Straße 38, 67056 Ludwigshafen am Rhein, Germany
niklas.loges@basf.com

[2] Simulation Science Center Clausthal-Göttingen, Clausthal University of Technology, Arnold-Sommerfeld-Straße 6, 38678 Clausthal-Zellerfeld, Germany
alexander.herzog@tu-clausthal.de

Abstract. Quality Control (QC) is essential for ensuring product quality and safety in food and healthcare/pharmaceutical industries. This paper presents the development and applications of the logistic processes digital twin at the QC laboratory at BASF Ludwigshafen. The digital twin is realized as a stochastic, event-driven simulation model that maps the laboratory's logistical processes including staffing, sample splitting, routing and joining, and prioritization. The model enables the identification of bottlenecks, estimation of staffing requirements, and prediction of sample turnaround times under various scenarios. The simulation, developed in collaboration with the Simulation Science Center Clausthal-Göttingen (SWZ), leverages the open-source tool "Warteschlangensimulator" and is validated against real operational data from 2021 and 2022. The results demonstrate the model's effectiveness in supporting decision-making for laboratory reorganization, staff upskilling, and workload management, including the integration of new product types. Future work aims to integrate the model into ongoing reporting systems and expand its scope to all QC laboratories at BASF Ludwigshafen.

Keywords: Digital Twin · Quality Control · Event-Driven Stochastic Simulation · Network Optimization

1 Introduction

Quality Control (QC) laboratories are pivotal in the assurance of product quality and regulatory compliance in the food, nutrition, and pharmaceutical industries, [EU05, ICH00]. Their responsibilities span from the release of raw materials and finished products, execution of stability studies, and adherence to Good Manufacturing Practice (GMP) requirements. The increasing complexity of laboratory operations, driven by a growing diversity of product portfolios, fluctuating workloads, and stringent regulatory demands, necessitates advanced tools for process optimization and decision support.

N. Merkert and A. Herzog (Eds.): SimScience 2025, CCIS 3027, pp. 27–39, 2026.
https://doi.org/10.1007/978-3-032-29720-4_3

Traditional approaches to laboratory management often rely on static reporting and retrospective analysis, which are insufficient for proactively identifying bottlenecks or forecasting the impact of operational changes. In this context, the concept of a "digital twin" – a dynamic, virtual representation of the laboratory's logistical processes – offers a transformative approach, see [Cre23, Soo23, Tao22]. By simulating the flow of samples, personnel allocation, and resource utilization, a digital twin enables laboratory managers to evaluate the consequences of process modifications, staffing changes, and new product introductions before implementation.

This paper details the conception, development, and application of a digital twin for the QC laboratory at BASF Ludwigshafen, focusing on its methodological foundation, model validation, and practical impact on laboratory operations.

2 Project Objectives

The "Digital QC Twin" project[1] was initiated with the following core objectives:

- **Bottleneck Identification:** Systematically detection and quantification of process bottlenecks and dependencies within the laboratory workflow, enabling targeted interventions.
- **Personnel Planning:** Accurately estimating staffing requirements for different analysis stations and sample types, supporting both short-term scheduling and long-term workforce development.
- **Turnaround Time Prediction:** Forecasting sample turnaround times under varying scenarios, including changes in workload, staffing levels, and process configurations.
- **Cost-Optimal Operation:** Facilitate the determination of cost-optimal operating points by balancing personnel deployment against throughput and turnaround time targets.
- **Continuous Adaptation:** Ensure that the simulation model remains adaptable to evolving operational conditions, such as shifts in product mix, regulatory requirements, or laboratory organization.
- **Integration with Reporting Systems:** Enable seamless integration of the simulation tool with existing reporting and data analytics platforms (e.g., PowerBI, see [PBI21]), fostering data-driven decision-making across the organization.

3 Methodology

3.1 Modeling Approach

The digital twin is constructed as a stochastic, event-driven simulation model, focusing on the logistical (rather than chemical or physical) aspects of laboratory operations [Bra18, Sch24]. The modeling approach is inspired by business process modeling [AS04, OMG07] and queueing theory [Bha08, Bro12, Sho18], representing the laboratory workflow as a directed flowchart consisting of more than 350 elements. (A simplified version of the flowchart will be presented in Sect. 4.1, see Fig. 1.) Each element of the flowchart corresponds to an organizational process step – such as sample reception, analysis, or approval – rather than a specific physical device.

[1] https://www.simzentrum.de/en/research/projects/basf

Key features of the modeling approach include:

- **Stochastic Representation:** Both sample arrivals and service times are modeled via random variables, typically using exponential, triangular, or log-normal probability distributions. For fitting the probability distributions to the historical data, mean, standard deviation, minimum and maximum of the histograms were calculated, and a distribution was fitted using the domain specific information on the actual process, e.g. exponential distribution of memoryless arrival streams, see [Sap96].
- This reflects the inherent variability in laboratory operations and enables robust scenario analysis.
- **Event-Driven Simulation:** The model operates on an event-driven basis, capturing the dynamic interplay between sample arrivals, processing, and resource allocation, see [Law13].
- **Abstraction for Flexibility:** The model deliberately abstracts from unnecessary detail, focusing on the key drivers of system performance. This ensures flexibility for future investigations and minimizes the data collection burden. It makes the model experimental and thus allows to extrapolate to investigate future questions. On the other hand, the abstraction has no negative effects on the accuracy of results if the removed details are selected carefully. For example, the arrival process is modeled via easily adjustable probability distributions instead of individual (historical) arrival times for each sample.
- **Scenario Experimentation:** The model is designed primarily for forward-looking scenario analysis, allowing laboratory managers to evaluate the impact of hypothetical changes (e.g., new product introductions, staffing adjustments) without reconstructing historical events.

3.2 Simulation Tool

The simulation is implemented using "Warteschlangensimulator",[2] a Java-based open-source tool developed at SWZ. This tool offers several advantages:

- **Graphical user-interface:** Modeling is performed using an easy-to-use graphical user-interface that does not require any programming skills to create models.
- **Open-Source and Extensible:** The simulator is freely available under an Apache 2.0 license, ensuring long-term sustainability and adaptability for BASF.
- **Interactive and Automated Modes:** It supports both interactive desktop use and integration with automated optimization and reporting systems. While animations are useful to understand, validate and present the model behavior, the fast simulation model allows to check many complex scenarios in a short period of time.
- **Open Interfaces:** Documented interfaces (XML file for models; direct loading and storing table data in file formats for spreadsheet processing tools) allow for seamless exchange of model parameters and statistical data with other systems, such as PowerBI dashboards.
- **Comprehensive Documentation:** The tool is accompanied by a bilingual (German/English) user-interface, bilingual online help, and documentation facilitating user training and adoption. A textbook about Warteschlangensimulator is also available, see [Her21].

[2] https://a-herzog.github.io/Warteschlangensimulator/

3.3 Data Collection, Model Calibration and Verification

A significant challenge in model development was the limited availability of detailed process data in existing IT systems. The available data recording was intended for reporting purposes, not to build digital twins based on them. To address this problem, the project adopted an iterative calibration approach, see [Rab08]:

- **Parameter Estimation:** Input parameters – such as sample arrival rates, staff availability, and service times – were estimated from a combination of historical records, aggregated statistics, and expert interviews with laboratory personnel.
- **Iterative Fitting:** The model was iteratively adjusted to match observed lead times and throughput data from 2021 (model training data), with subsequent validation against 2022 data (model testing data). Between 2021 and 2022 there have been no structural changes in the laboratory. Each year shows the same typical pattern in terms of the number of arrivals per sample type and month, and in the employee availability per month (due to vacation, etc.). So, it was a natural decision to split the available data by year and use one of them for training and one for testing.
- **Expert Knowledge Integration:** The detailed knowledge of laboratory staff was essential for accurately capturing process nuances, such as sample routing rules, shift schedules, and special handling for certain sample types.

4 Simulation Model

4.1 Process Flow and Model Structure

The simulation model represents the end-to-end journey of the samples through the QC laboratory, from initial arrival to final approval. The process flow is structured as follows (see Fig. 1):

- **Sample Arrival:** Samples arrive at a central source station, characterized by product-specific inter-arrival times, see Table 1. The model considers eight main product groups and four product clusters (consisting of multiple smaller product groups each), each with distinct arrival patterns. Since the samples arrive independently of each other, the exponential distribution was selected to describe the arrival process.
- **Sample Reception:** Upon arrival, samples are routed to one of five sample reception stations (PA-0 to PA-4), each with its own processing characteristics. Reception includes optional delays, processing, and mandatory quality control checks.
- **Sample Splitting and Routing:** Depending on product type, samples are split into sub-samples and routed to a subset of 15 analysis stations ("boxes"). Each box represents a specific analytical process and requires staff with appropriate qualifications.
- **Analysis and Approval:** Sub-samples are processed at their assigned boxes, passing through preprocessing, main processing, and control stages. Once all sub-sample tests for a given sample are complete, the results are consolidated, and the sample is approved for release.
- **Shift Schedules and Working Hours:** The model incorporates a simplified shift schedule (8 active hours per day), excludes weekends and holidays, and adjusts throughput calculations accordingly.

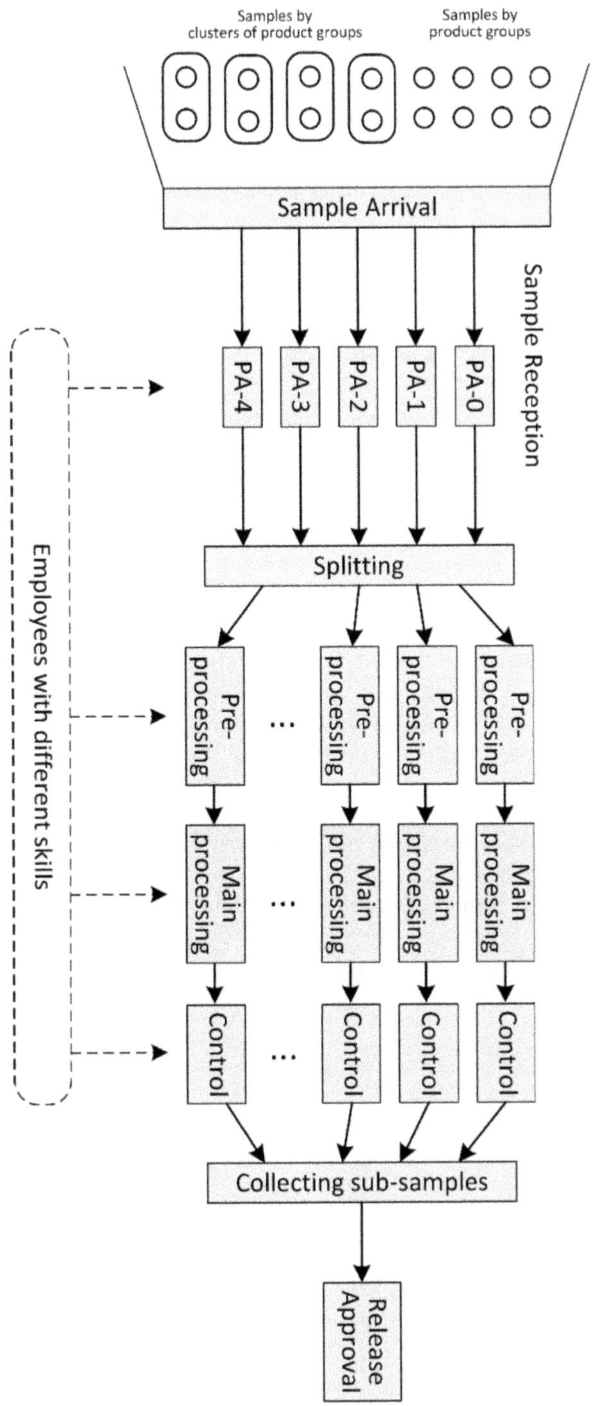

Fig. 1. Structure of the developed digital twin model.

- **Statistical Monitoring:** Throughout the simulation, key performance indicators – such as waiting times, processing times, lead times, and resource utilization – are tracked for each station and product type.

4.2 Model Abstraction and Experimentation

A central design principle is the abstraction of real-world complexity to enable efficient experimentation. For example:

- **Aggregated Parameters:** Rather than modeling times of every individual sample, the model uses aggregated arrival and service time distributions, enabling rapid evaluation of future scenarios.
- **Flexible Routing:** The routing logic is parameterized, allowing easy modification to reflect changes in product mix or laboratory organization (like employee training).
- **Scenario Scripting:** Custom scripts can be used to filter and analyze simulation results, supporting direct comparison with historical data and facilitating integration with external analytics tools.

4.3 Model Parameters

The complete model parameters include both the shift schedules of individual employees and BASF's production specifics and therefore cannot be made fully available to the public. Table 1 shows the total arrivals per sample type per year and the assigned sample reception station for each sample type. The actual inter-arrival times per sample type

Table 1. Number of sample arrivals per year and sample type to reception station assignments

Sample type	Number of arrivals per year		Assigned sample reception stations
	2021	2022	
Product A	778	448	PA-4
Product B	247	165	PA-3
Product C	144	112	PA-3
Product D	274	222	PA-2
Product E	144	127	PA-4
Product F	263	322	PA-3
Product G	880	783	PA-3
Product H	206	158	PA-0
Sample cluster 1	905	1025	PA-3
Sample cluster 2	1732	1070	PA-3
Sample cluster 3	1884	1493	PA-3
Sample cluster 4	623	1053	PA-1

Table 2. Processing and control times at the sample reception stations

Reception station	Processing times	Control
PA-0	0 min	Triangular [0;1;2] minutes
PA-1	Triangular [0;30;60] minutes	
PA-2	30 min + triangular [35;45;50] minutes	
PA-3	30 min + triangular [20;25;30] minutes	
PA-4	30 min + triangular [20;40;60] minutes	

used in the simulation are modeled using the exponential distribution since the arrivals are independent. The mean inter-arrival times (in minutes) are calculated by dividing $365 \cdot 1440$ by the total number of arrivals per year.

The processing times at the reception stations are subject to the triangular distribution, see Table 2. There are multiple employees at the sample reception. Each employee can work at each sample reception station. The processing at the main station is (in the base scenario) is carried out by a single employee in some cases. Therefore, the processing times used at the main stations cannot be published due to employee data protection reasons.

4.4 Model Validation

Validation was performed by comparing simulation statistic outputs with real operational data from 2021 and 2022. Key input parameters – such as sample arrival rates, staff availability, and service times – were derived from 2021 historical records and expert interviews. The validation process involved:

- **Calibration:** Model parameters were tuned to match observed lead times and throughput for 2021, using aggregated statistics for each product group and analysis station. The comparison focused on three main performance indicators: (1) waiting and processing times at the sample reception, (2) waiting and processing times at the main analysis stations ("boxes"), and (3) total sample turnaround times.
- **Cross-Validation:** The calibrated model was then applied to 2022 data, with only arrival rates updated to reflect the changes in workload. The resulting lead times and throughput were compared to the actual historical values, demonstrating good consistency – typically within 5% for key performance indicators.
- **Sensitivity Analysis:** The model's robustness was further tested by varying key parameters and assessing the impact on system performance, confirming its suitability for scenario-based decision support.

The validation results demonstrated a high degree of consistency between simulation and real-world results, particularly for the critical analysis stages. For example, the average deviation in total turnaround times was less than 5% for most product types, confirming the model's suitability for both qualitative and quantitative scenario analysis.

5 Results

5.1 Bottleneck and Throughput Analysis

The first major objective of the digital twin was to identify bottlenecks within the laboratory workflow. The model revealed that, for most sample types, the overall turnaround time is determined by the slowest analysis station (which also has the highest utilization rate) encountered by each sample – a classic manifestation of the "Theory of Constraints", see [Gol90]. Specifically, the stations labeled MIK (Microbiological testing), UV (UV spectroscopy), and PK (Physical properties testing) were found to be the most frequent bottlenecks across a wide range of product types.

The analysis further decomposed the time spent at these bottlenecks into three components: (1) preprocessing delays, (2) waiting times for available staff, and (3) actual processing times. For instance, at the MIK station, preprocessing accounted for 95% of the total time, while at UV and PK, waiting times were the dominant factor (82% and 72%, respectively). This decomposition provides actionable insights: reducing preprocessing at MIK or increasing staffing flexibility at UV and PK can yield substantial improvements in overall laboratory throughput.

5.2 Scenario-Based Experiments

Introduction of New Product Types

The model was used to simulate the impact of introducing a new product type, "Kollidon 90 Evo", which would add 60 additional samples over a four-week period. Two scenarios were tested: (a) all additional samples arriving in a short burst, and (b) arrivals distributed evenly over the period. The results showed that concentrated arrivals led to significant temporary increases in turnaround times, indicating that the system operates near capacity limit under normal conditions. However, when arrivals were spread out, see Fig. 2, the system absorbed the additional workload with acceptable increases in average turnaround times (peaks not much higher than the peaks caused by the stochastic nature of the arrivals and the service itself). Quantitatively, the additional workload resulted in a 10.6% increase in sample reception and a 3.0% increase in analysis workload per week, both of which were manageable under the distributed scenario.

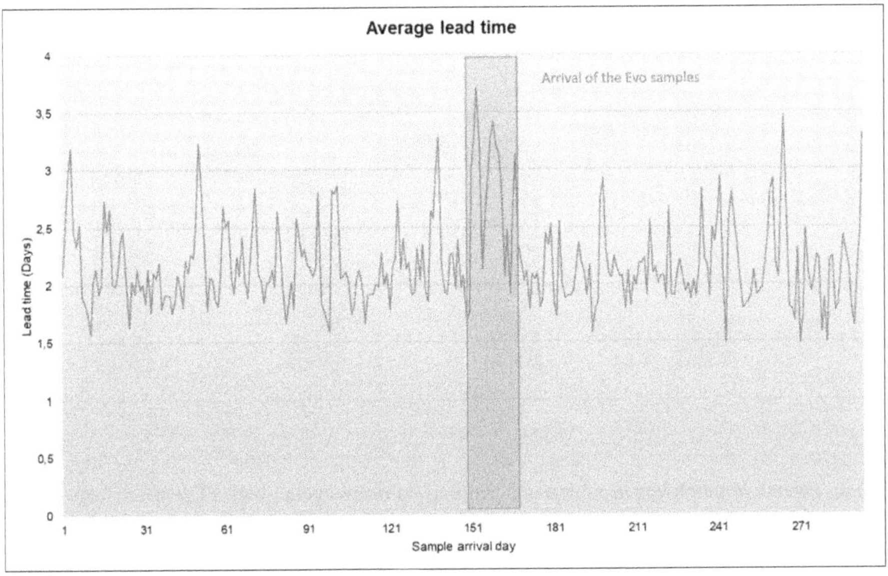

Fig. 2. Effect of the arrival of the new "Kollidon 90 Evo" samples on the overall average lead times.

Staff Upskilling and Flexibility

The simulated models explored the effects of cross-training staff to operate multiple analysis stations. Four variants were compared:

- **Baseline scenario:** Each employee is only trained to operate at one analysis station.
- **Variant 1:** Selected staff received additional qualifications for specific stations.
- **Variant 2:** Staff were reorganized into fewer, larger groups with broader qualifications.
- **Variant 3:** All staff were qualified for all stations.

The results were striking: Variant 1 reduced average turnaround times by 11%, Variant 2 by 22%, and Variant 3 by 24% compared to the baseline, see Fig. 3. Notably, the largest gains were achieved by moving from highly specialized (baseline) to moderately flexible staffing (variants 1 and 2); further increases in flexibility (variant 3) yielded diminishing returns. This suggests that targeted upskilling, rather than universal cross-training, is the most cost-effective strategy for improving laboratory performance.

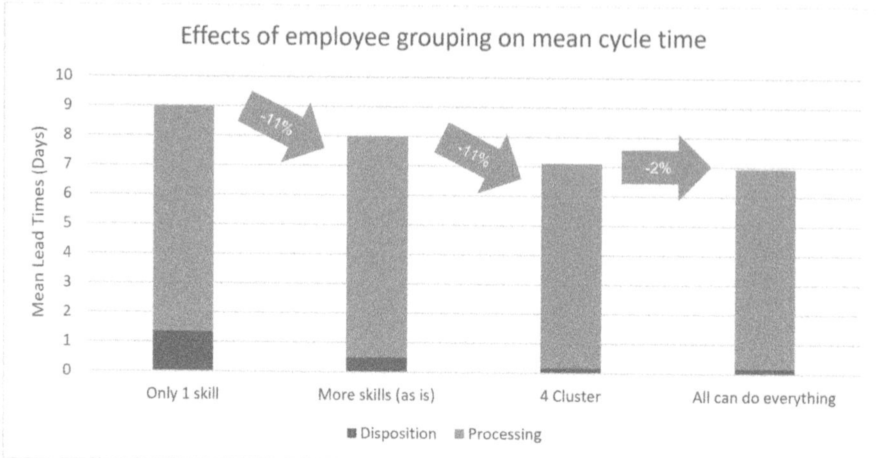

Fig. 3. Effects of employee grouping and training on mean cycle times. (The percentages in the green arrows represent the percentage change from one step to the next.)

Impact of Staff Vacations

The model also quantified the effects of staff vacations on laboratory performance. Simulations showed that overlapping vacations among key staff groups led to significant backlogs and increased turnaround times. However, the introduction of targeted mitigation measures – such as temporary staff or upskilling existing staff for critical roles – substantially dampened these negative effects. For example, qualifying two additional staff members for sample reception during peak vacation periods nearly eliminated the backlog, demonstrating the value of strategic flexibility in workforce planning.

5.3 Quantitative Summary

The following Table 3 summarizes key findings from the simulation experiments:

Table 3. Quantitative Summary of the different scenarios.

Scenario	Change in average turn-around time or change in WIP	Notes
Baseline (2022)	0%	Reference scenario
Variant 1 (targeted upskilling)	Turn-around time – 11%	Selected staff gain extra qualifications
Variant 2 (group reorganization)	Turn-around time – 22%	Fewer, larger, more flexible staff groups
Variant 3 (full flexibility)	Turn-around time – 24%	All staff qualified for all stations

(*continued*)

Table 3. (*continued*)

Scenario	Change in average turn-around time or change in WIP	Notes
New product (distributed)	Turn-around time + 3% (boxes), +10.6% (reception)	Manageable with distributed arrivals
New product (burst)	Turn-around time + 20% (temporary spikes)	System overload if arrivals are clustered
Vacation (no mitigation)	Significant backlog WIP +900% for 4 weeks	Backlog persists for weeks post-vacation
Vacation (mitigation)	Minimal backlog WIP +100% for 2 weeks	Targeted upskilling/temporary staff works

These results demonstrate the power of the digital twin approach for laboratory logistics: it enables data-driven decision-making, supports proactive workforce planning, and provides a robust platform offering reliable predictions for continuous process optimization.

6 Discussion

The implementation of a digital twin for the QC laboratory at BASF Ludwigshafen has yielded significant insights and practical benefits:

- **Data-Driven Decision Support:** The simulation model provides laboratory managers with a powerful tool for evaluating the impact of operational changes before implementation. By simulating various scenarios – such as new product introductions, staffing adjustments, or process reorganizations – decision-makers can anticipate bottlenecks, quantify resource requirements, and optimize turnaround times.
- **Workforce Planning:** The model's ability to simulate staff upskilling and flexible deployment has informed strategic decisions regarding training investments and personnel allocation. The finding that moderate increases in staff flexibility yield substantial performance gains, while universal cross-training offers diminishing returns, is particularly valuable for cost-effective workforce development.
- **Operational Resilience:** The model's capacity to simulate the effects of staff vacations and other disruptions has enabled proactive planning for peak periods and contingency scenarios. Targeted mitigation measures – such as temporary staff or selective upskilling – can be evaluated and implemented with confidence.
- **Continuous Improvement:** The digital twin serves as a living model, continuously updated to reflect changes in laboratory operations, product mix, and organizational structure. Its integration with reporting and analytics platforms ensures that it remains a central component of the laboratory's continuous improvement strategy.

7 Conclusion and Outlook

The digital twin of the QC laboratory at BASF Ludwigshafen exemplifies the transformative potential of simulation-based approaches for logistically process optimization in complex laboratory environments. By providing a dynamic, data-driven platform for scenario analysis and decision support, the model has enabled significant improvements in operational efficiency, workforce planning, and organizational resilience.

Future work in the project will focus on:

- **Integration with Live Reporting:** Embedding the simulation model within ongoing reporting and analytics systems (e.g., PowerBI) to enable real-time monitoring and decision support.
- **Expansion to Additional Laboratories:** Extending the model's scope to encompass all QC laboratories at BASF Ludwigshafen, creating a comprehensive digital twin ecosystem.
- **Advanced Analytics and Optimization:** Leveraging the model for advanced analytics, such as predictive maintenance, automated scheduling, and optimization of laboratory workflows.
- **Blueprint for Industry:** The approach and methodology developed in this project serve as a blueprint for similar initiatives in other industrial laboratory settings, demonstrating the value of digital twins for operational excellence.

References

1. Aguilar-Savén, R.S.: Business process modelling: review and framework. Int. J. Prod. Econ. **90**(2), 129–149 (2004)
2. Bhat, U.N.: An Introduction to Queueing Theory: Modeling and Analysis in Applications. Birkhäuser, Boston (2008)
3. Bracht, U., Geckler, D., Wenzel, S.: Digitale Fabrik. Springer, Heidelberg (2018)
4. Borovkov, A.: Stochastic Processes in Queueing Theory, vol. 4. Springer Science & Business Media, New York (2012)
5. Crespi, N., Drobot, A.T., Minerva, R.: The Digital Twin: What and why?. The Digital Twin, pp. 3–20. Springer International Publishing, Cham (2023)
6. EU Guidelines to Good Manufacturing Practice Medicinal Products for Human and Veterinary Use Part I Chapter 6 Quality Control (2005). https://health.ec.europa.eu/document/download/f0a9c113-d316-4f33-8633-9275b007e459_en?filename=2005_10_chapter_6_en.pdf
7. Goldratt, E.M.: Theory of Constraints. North River, Croton-on-Hudson (1990)
8. Herzog, A.: Simulation mit dem Warteschlangensimulator – Mathematische Modellierung und Simulation von Produktions- und Logistikprozessen. Springer, Heidelberg (2021)
9. ICH Harmonised Tripartite Guideline: Good Manufacturing Practice Guide for active Pharmaceutical Ingredients. https://ich.org/page/quality-guidelines
10. Law, A.M.: Simulation Modeling and Analysis, 5th edn. McGraw Hill, Columbus (2013)
11. Specification: OMG Available. Omg unified modeling language (omg uml), superstructure, v2. 1.2. Object Management Group 70 (2007)
12. Power, B.I., et al.: Microsoft power bi. (2021). https://powerbi.microsoft.com/en-us
13. Rabe, M., Spieckermann, S., Wenzel, S.: Verifikation und Validierung für die Simulation in Produktion und Logistik: Vorgehensmodelle und Techniken. Springer, Heidelberg (2008)
14. Sapsford, R., Jupp, V.: Data collection and analysis. Sage. London. (1996)
15. Schluse, M.: Experimentierbare Digitale Zwillinge. Konvergenz von Simulation und Realität, Springer, Heidelberg (2024)
16. Shortle, J.F., Thompson, J.M., Gross, D., Harris, C.M.: Fundamentals of Queueing Theory. John Wiley & Sons, Hoboken (2018)

17. Soori, M., Arezoo, B., Dastres, R.: Digital twin for smart manufacturing, a review. Sustain. Manufact. Serv. Econ. **2**, 100017 (2023)
18. Tao, F., Xiao, B., Qi, Q., Cheng, J., Ji, P.: Digital twin modeling. J. Manuf. Syst. **64**, 372–389 (2022)

Wettability of An Aluminum Droplet on An Alumina Substrate

Shan Lyu(✉), Ganesh Raj Ramakrishnan, and Nina Merkert

Institute of Metallurgy, Technical University of Clausthal, Clausthal-Zellerfeld, Germany

shan.lyu@tu-clausthal.de

Abstract. We simulate the wettability of an aluminum (Al) droplet on an alumina (α-Al_2O_3) via the molecular dynamics (MD) method. We consider the chemical reaction in this system, i.e. the break and formation of bonds. During the wetting process, liquid Al droplet spreads over the α-Al_2O_3 substrate until it reaches its equilibrium state, forming a stable contact angle. The important parameters are (1) the size of the droplet; (2) the system temperature; and (3) the crystallographic orientation of α-Al_2O_3. The findings reveal that the influence of droplet size on the wettability is limited to a threshold. With elevated temperatures, the chemical interaction and wettability of the droplet is enhanced. The crystallographic orientation of the substrate affects the new Al-O bond formation and thus the oxidation and wetting process.

Keywords: Wettability · molecular dynamics · α-Al_2O_3 · Oxides

1 Introduction

Recent advances in transportation technology require lightweight materials with a high power-to-weight ratio to improve fuel efficiency and durability. This involves the use of aluminum-based alternatives. Thus, extensive research has been directed towards understanding the wettability of pure aluminum (Al) on an alumina (α-Al_2O_3) substrate [1].

There has been a substantial number of numerical works focusing on studying the wettability of Al droplets on α-Al_2O_3 substrates. Hong et al. [2], through reactive force field (ReaxFF)-based MD simulations, studied the oxidation kinetics of Al nanoparticles, identifying hotspots and void formation mechanisms as crucial to oxide growth. They observed that localized hotspots substantially enhance oxygen diffusion, promoting rapid oxide formation and affecting the thickness and structure of the oxide layer, critically influencing wetting and adhesion at the Al–Al_2O_3 interface. Yousefi et al. [3] conducted MD simulations using the CTI + Embedded Atom Model (EAM) potential to examine the surface tension of Al-O systems and the effects of oxygen in the wetting process. They found that the surface tension decreases significantly as the oxygen ratio increases until a saturation point is reached. Tahani et al. [4] employed MD

N. Merkert and A. Herzog (Eds.): SimScience 2025, CCIS 3027, pp. 40–54, 2026.
https://doi.org/10.1007/978-3-032-29720-4_4

simulations using COMB3, Morse, and Tersoff potentials to study the diffusion phenomenon at the Al–Al_2O_3/AlSi12 interface. The results indicated significant interdiffusion at the interface, with aluminum atoms in the AlSi12 alloy showing notably higher mobility compared to those within the Al_2O_3. They further demonstrated that the diffusion process was largely independent of the substrate termination type of Al_2O_3. Zhuo et al. [5] further investigated the interfacial dynamics by simulating the orientation-dependent wetting behavior of Al droplets on different α-Al_2O_3 crystal planes using the EAM and Lennard-Jones (L-J) potentials. The results concluded that c-plane (0001) exhibited the strongest adhesion and lowest wetting angle, illustrating how crystallographic orientation impacts oxide layer interactions.

Parallel to the numerical analyses, extensive experimental studies [6–8] of Al/Al_2O_3 can be found. Shen et al. [6] found the negligible influence of the system temperature on the initial equilibrium contact angle. However the equilibrium contact angle varies significantly with the substrate orientation. Bao et al. [7], performed wetting experiments under different atmospheric conditions, such as vacuum and argon environments. They observed that oxide removal in vacuum significantly reduced the contact angle from $115°$ to $60°$, emphasizing the critical role of oxide layers. Shen et al. [8], highlighted the major effects of oxide layer thickness on the contact angle, which confirmed that efficient oxide removal is essential for achieving better wettability.

In the current paper, we re-visit the Al/αAl_2O_3 system using MD simulations combined with ReaxFF potential, which inherently accounts for bond breaking and formation during interfacial interactions. A comprehensive analysis is performed to understand the effects of system temperature, droplet size, and substrate orientation on the wetting behavior of the Al droplet. These parameters are of particular importance, as they govern the performance of Al in applications such as joining, coatings, and high-temperature structural materials. The present study offers a systematic and atomistic-scale exploration over a wide range of previously unaddressed conditions, enabling direct correlation between atomic-scale reaction mechanisms and macroscopic wetting phenomena. Furthermore, by quantitatively comparing different surface orientations, this work provides insights into the dynamic wettability of reactive liquid metals, which is rarely covered in earlier studies.

2 Model and Simulation Method

In this work, we perform MD simulations using the Large-scale Atomic Massively Parallel Simulator (LAMMPS). The ReaxFF potential [2] is used to describe the Al/α-Al_2O_3, in which the pairwise interactions of Al-Al and Al-O are parameterized. ReaxFF allows bond to form and break dynamically and the total energy (E_{total}) of the system is a sum of many contributions:

$$\begin{aligned} E_{\text{total}} &= E_{\text{bond}} + E_{\text{over/undercoordination}} + E_{\text{valence}} \\ &\quad + E_{\text{torsion}} + E_{\text{conjugation}} + E_{\text{vdW}} + E_{\text{Coulomb}}. \end{aligned} \tag{1}$$

where E_{bond} is the bond energy accounting for bond stretching and breaking, $E_{\text{over/undercoordination}}$ the penalty term for correcting atoms with unreasonable bond numbers, E_{valence} the valence angle energy accounting for bond angles based on coordination, E_{torsion} the torsional energy describing dihedral rotations, $E_{\text{conjugation}}$ the conjugation energy, E_{vdW} the van der Walls interactions counting short- and long-range dispersion forces, E_{Coulomb} the coulomb interactions that computes partial charges dynamically. Figure 1 shows the melting temperatures of the pure Al slab and α-Al_2O_3 using the ReaxFF potential. The simulated melting points are in reasonable agreement with the experiments, i.e. ($T_{sim}^{Al} = 997$ K and $T_{sim}^{\alpha-Al_2O_3} = 2663$ K; $T_{exp}^{Al} = 933$ K and $T_{exp}^{\alpha-Al_2O_3} = 2345$ K). Furthermore, the adhesive energy of a single oxygen atom on the Al(111) surface at different sites (namely HCP and FCC sites) are compared with that from the Ab-initio method [9], i.e. for HCP: $E_{sim} = 8.46$ eV and $E_{ab-initio} = 8.05$ eV; for FCC:$E_{sim} =$ 8.04 eV and $E_{ab-initio} = 8.31$ eV. The good agreement suggests the capability of the ReaxFF potential in modeling the wetting process of the Al/α-Al_2O_3 system.

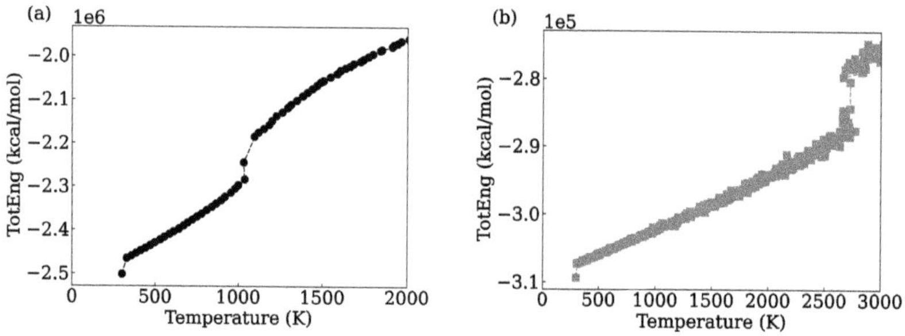

Fig. 1. Calculation of melting temperatures of Al and α-Al_2O_3 using ReaxFF potential.

Figure 2 illustrates the initial configuration of the Al/α-Al_2O_3 system. Prior to the wetting process, the Al droplet and the α-Al_2O_3 are equilibrated separately at their respective target temperatures. For this purpose, we first perform an NPT heating process (constant number of particles, temperature, and pressure) with a time step of 0.25 fs for 200 ps, followed by an NVT equilibration (constant number of particles, volume, and temperature) for an additional 200 ps. The systems are considered equilibrated once their temperature and total energy fluctuate around stable mean values without any systematic drift. The substrate is well equilibrated with the following criteria: the temperature fluctuates around 880 K with variations of 10 K, while the total energy remains nearly constant apart from small thermal fluctuations ($\leq$0.2%). The bottom layers (2 Å) of α-Al_2O_3 are fixed to avoid vertical atom shift. A thermostat zone of 18 Å is controlled using the NVT ensemble. The top layers (13 Å) are allowed to move freely using the NVE ensemble (constant number of particles, volume, and energy) to present artificial thermostatting. The Al droplet is positioned 4 Å above the α-Al_2O_3 surface before the wetting process begins. Periodic boundary conditions are applied in the x and y directions, while a fixed boundary condition is employed in the z direction. The wetting system is maintained at a constant temperature for 600 ps, during which the wettability of the Al droplet are analyzed.

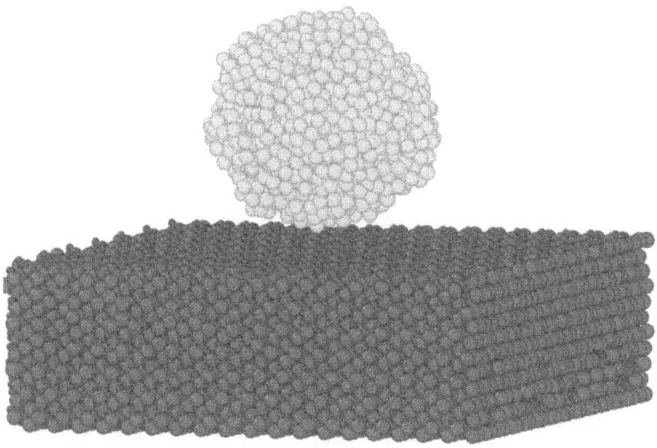

Fig. 2. Initial configuration of the Al/α-Al_2O_3 system. Atoms marked in yellow are Al droplet, consisting of 1481 atoms. The α-Al_2O_3 substrate contains 27930 atoms, with red atom being Al and blue atom being Oxygen (O). The system temperature is 1000 K (Color figure online).

To determine the wettability of the Al droplet, we assess the evolution of the contact angle. In the Al/α-Al_2O_3 wetting system, oxidation and mutual diffusion are expected to occur. Therefore, it is impractical to locate the droplet boundary based on their atomic types, i.e. convex hull method [10] is no longer ideal. Instead, we extract the droplet shape in a two-dimensional xz slice and apply a spatial-binning method.

To extract the contact angle, we use only boundary points, i.e. $(x_k, z_{\max}(x_k)) \in \mathcal{B}$ with $x_k \in [x_1, x_2]$, and fit them by least squares to a circular arc,

$$(x - x_0)^2 + (z - z_0)^2 = R^2,$$

with fit parameters x_0, z_0 and R. The substrate is taken as the plane $z = z_s$. The fitted circle intersects the substrate at the contact point (x_c, z_s), and the contact angle θ is obtained from the local tangent to the circle at this point. The outward normal of the circle at (x_c, z_s) is

$$\mathbf{n} = \frac{1}{R}\,(x_c - x_0,\ z_s - z_0),$$

so that the angle between the droplet interface and the substrate is

$$\theta = \arccos\left(\frac{z_s - z_0}{R}\right),$$

measured through the liquid phase. Repeating this procedure for each slice yields the time evolution of the droplet contact angle.

Bond order is computed internally by ReaxFF as a continuous, distance dependent quantity. We obtain the bond number (BO) from the ReaxFF calculation and classify a pair as bonded if BO exceeds the cutoff bond number (BO_{cut}). The cutoff values BO_{cut} are chosen from the peaks of the bond-order distributions in Fig. 8, which cleanly separate bonded interactions from the low-bond-order tails. Disruption of bonding is then quantified as the loss of such bonds (decrease in the number of pairs with $BO > BO_{\mathrm{cut}}$) as a function of time.

3 Results and Discussion

In this work, we investigate the influence of droplet size, system temperature and the substrate orientation on the wettability of Al droplets. Four radii of Al droplets (namely 10 Å, 15 Å, 18 Å, and 20 Å) and four system temperatures (700 K, 900 K, 1000 K, and 1100 K) are considered. Four crystallographic orientations, as shown in Fig. 3, i.e. c-plane ((0001) Al-terminated), c-plane ((0001) O-terminated), a-plane (1–100), and r-plane (11–20), are tested. The choice of the parameter is based on the previous work [10]. We quantify the wettability by assessing the contact angle and the degree of oxidation by counting the newly formed Al-O bonds.

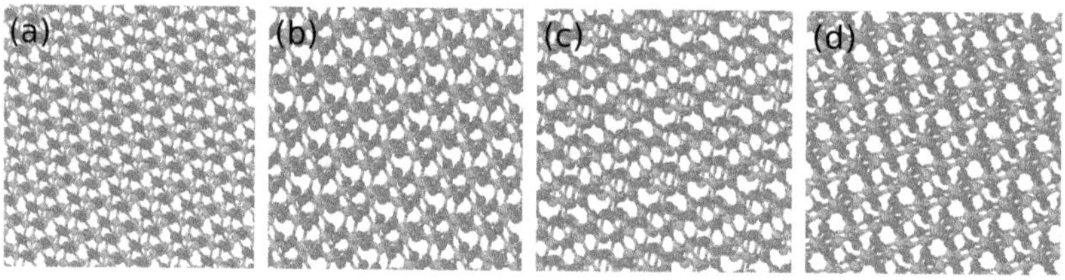

Fig. 3. Illustrations of α-Al_2O_3 substrates with different crystallographic orientations. (a) c-plane ((0001) Al-terminated), (b) c-plane ((0001) O-terminated), (c) a-plane (1–100), and (d) r-plane (11–20).

In a typical wetting process, the Al droplet gradually spreads across the α-Al_2O_3 substrate and increases its wetting area. The initial spherical Al droplet transitions into a semi-spherical shape, while the oxygen atoms from the substrate react with the Al droplet, resulting in the formation of new Al-oxides. At the equilibrium state, the contact angle is almost stable and the shape of the Al droplet remains unchanged.

3.1 Effect of Droplet Size

Figure 4 presents the evolution of Al droplets with different initial radii on the α-Al_2O_3 substrate, captured at 10 ps, 100 ps, and 500 ps. In all cases, the initially spherical Al droplets spread over time, transitioning toward a semi-spherical morphology and increasing their wetting area with the substrate until an equilibrium state is reached. For the smallest size droplet (10 Å), as shown in Fig. 4(a), spreading is most pronounced. The droplet spreads rapidly by 100 ps and forms an almost uniform thin layer by 500 ps. The droplet also undergoes extensive oxidation, with oxygen atoms from the α-Al_2O_3 substrate diffusing into the entire droplet and forming new Al-O bonds. Spontaneously, the Al-Al metallic bonding is significantly disrupted. For the size of 15 Åin Fig. 4(b), the droplet maintains a cap-like shape by 100 ps but spreads considerably by 500 ps. The oxidation proceeds efficiently. However, part of the Al droplet still persists at the intermediate stage before being consumed by the oxygen to form the Al-O bond. By further increasing the droplet size to 18 Å and 20 Å, as seen in Fig. 4(c-d), the droplet exhibits noticeably slower wetting kinetics. A considerable fraction of the Al atoms remain bonded metallically by 100 ps despite interfacial oxidation. A closer look at 500 ps shows that the oxidation mainly occurring at the interface and droplet surface. The droplet retains a semi-spherical shape at the equilibrium stage.

Figure 5 quantifies the evolution of the contact angle for the different Al droplet sizes over a time interval of 500 ps. The contact angle decreases significantly during the initial wetting stage and follows by a slower relaxation towards equilibrium. The smallest droplet spreads the fastest, with its contact angle rapidly falling below 30° within the first 100 ps and stabilizing around 20° by 200 ps. Larger droplets (e.g., 18 and 20 Å) exhibit more gradual spreading, resulting in higher contact angles during the early stages. However, all droplets converge to a similar value of approximately 23° at the equilibrium state, suggesting that while the equilibrium wettability is nearly size-independent, the spreading kinetics are strongly affected by the droplet size. The convergence of the equilibrium contact angle justifies the selection of a 18 Å radius for Al droplets in subsequent simulations, considering a balance between physical accuracy and computational efficiency.

3.2 Effect of System Temperature

Figure 6 illustrates the influence of system temperature on the wettability and oxidation behavior of an Al droplet on the α-Al_2O_3 substrate at 700 K, 900 K, and 1100 K, captured at 10 ps, 100 ps, and 600 ps. Noted that both the Al droplet and α-Al_2O_3 are kept at the same temperature. At 700 K, as shown in Fig. 6(a), the droplet maintains a relatively spherical morphology during the early stages, and although spreading is evident by 600 ps, the wetting kinetics remain slow. Oxidation occurs predominantly at the liquid-solid interface and a substantial metallic bonding persists due to limited oxygen diffusion at a such low temperature. At 900 K, the wetting behavior is promoted, as evident in Fig. 6(b), the

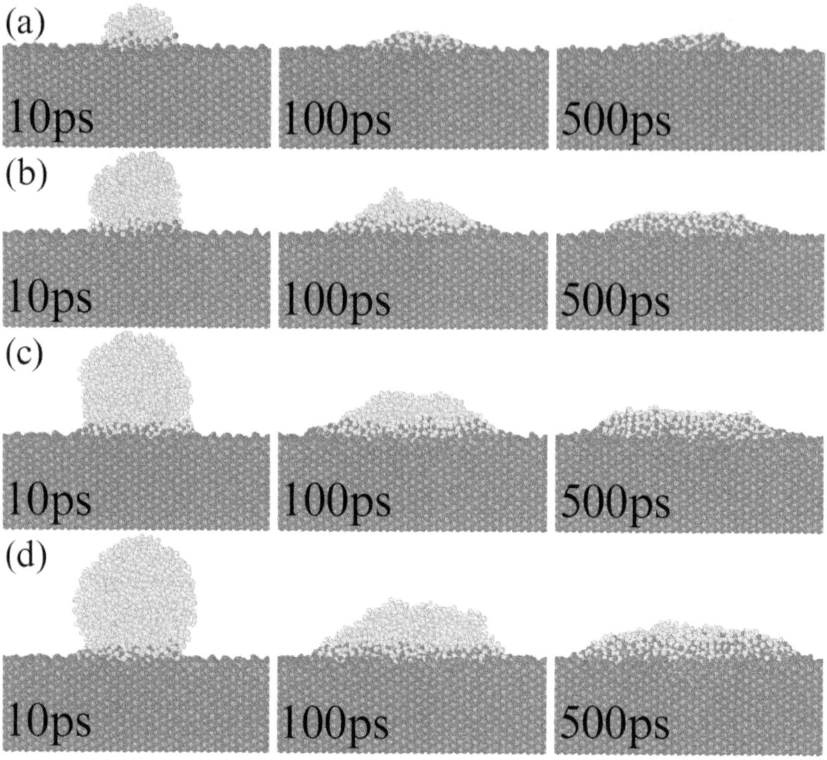

Fig. 4. Evolution of Al droplets on the α-Al_2O_3 substrate for different Al droplet size. The Al droplet is of the radius (a) 10 Å (259 atoms), (b) 15 Å (856 atoms), (c) 18 Å (1481 atoms), and (d) 20 Å (2015 atoms). The snapshots are taken at 10 ps, 100 ps, and 500 ps.

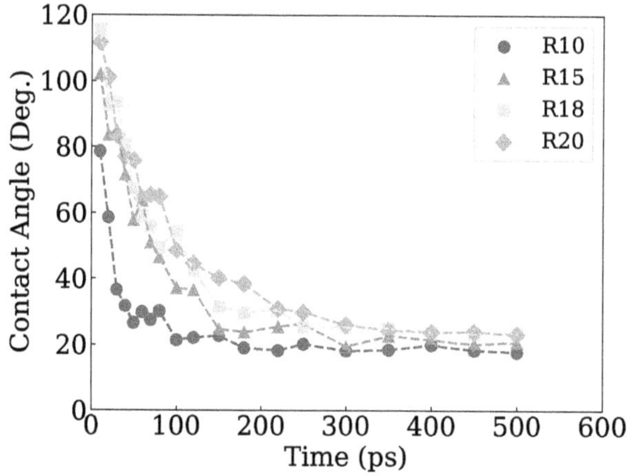

Fig. 5. Variation of contact angle as a function of droplet radius on the α-Al_2O_3 substrate at 1000 K.

droplet propagates more rapidly. At 600 ps, the contact area between the droplet and the substrate is much larger than that at 700 K. Furthermore, oxidation is more pronounced as the oxygen atoms diffusing deeper into the droplet. The Al–O bond formation is no longer limited to the interfacial layer. The metallic bonding within the Al droplet begins to diminish as the progress of the oxygen incorporation. At 1100 K, the droplet exhibits the fastest spreading behavior and the most extensive oxidation. Within the first 100 ps, the droplet already shows substantial spreading, and by 600 ps it forms a thin, flattened film-like structure across the substrate. The oxidation process is also significantly accelerated, as shown in Fig. 6(c), with a widespread disruption of the Al–Al metallic bonding and extensive formation of new Al-O bonds. At the equilibrium state, most of the droplet is oxidized and no clear boundary between the oxidized interface and the remaining droplet can be distinguished.

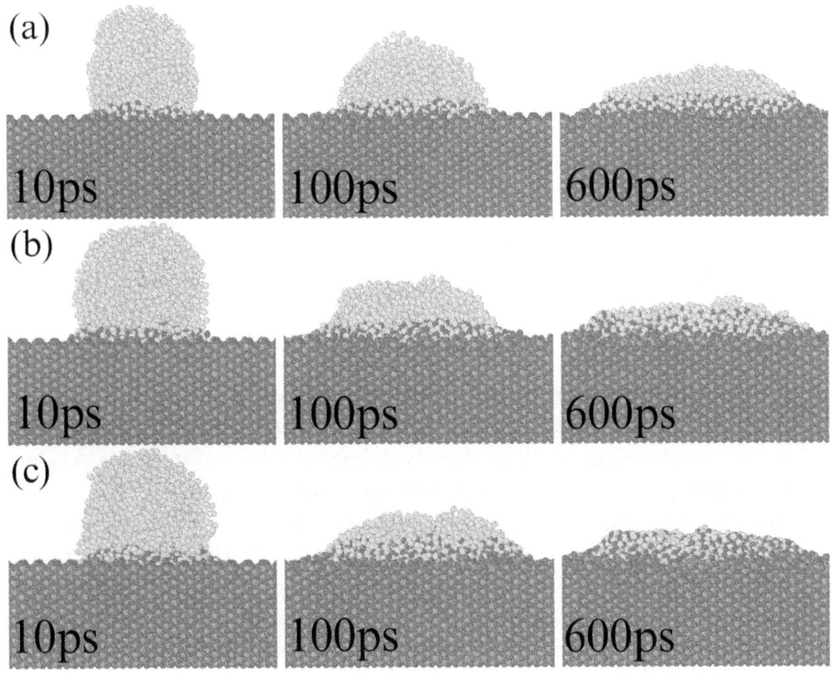

Fig. 6. Evolution of Al droplets on the α-Al_2O_3 substrate at different system temperatures. The Al droplet is of the radius 18. The temperatures are (a) 700 K, (b) 900 K, and (c) 1000 K. The snapshots are taken at 10 ps, 100 ps, and 600 ps.

Figure 7 shows the variation of contact angle as a function of time for Al droplets on the α-Al_2O_3 substrate at four different temperatures (700 K, 900 K, 1000 K, and 1100 K). The contact angle exhibits a clear dependence on the temperature. At 700 K, the droplet exhibits slow wetting kinetics, with the contact angle decreasing gradually and remaining relatively high (∼45°–50°). In contrast, at elevated temperatures (e.g., 900–1100 K), the contact angle decreases much more rapidly, dropping below 40° within the first 100 ps. At 1000 K and 1100 K, the droplets reach equilibrium values of ∼20°–25°, indicating enhanced wettability at higher temperatures.

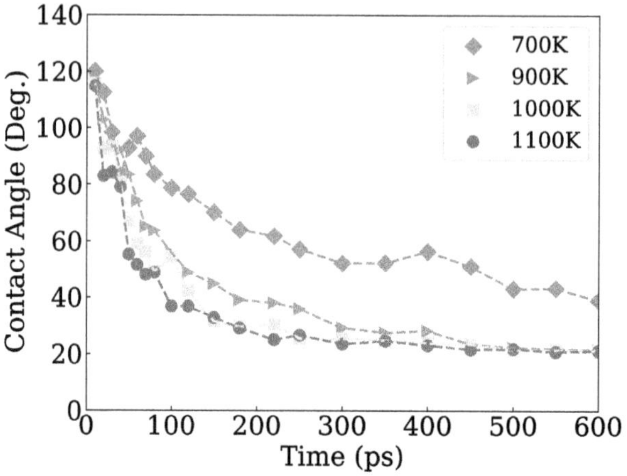

Fig. 7. Variation of contact angle as a function of system temperatures on the α-Al_2O_3 substrate.

Figure 8 presents the bond order distribution for different atomic pairs in the Al/α-Al_2O_3 system at 700 K, taken at 175 ps. This selected time is based on the criterion, for which oxidation has already occurred while part of the Al droplet still retains metallic bonding. Al(s) and Al(l) are Al atoms in the substrate and in the droplet, respectively. Fig. 8(a) shows the Al(s)–O bonds within the Al_2O_3 substrate. The peak is in the range of 0.6–0.7 that corresponds to stable Al-O coordination. Figure 8(b) demonstrates the Al(l)-O newly formed bonds, where the peak is similarly centered but slightly broader due to the less-ordered oxidation. Figure 8(c) corresponds to Al(l)-Al(l) metallic bonds, which has a peak around 0.5–0.6. Based on these distributions, cut-off distances are selected to distinguish bonded interactions from non-bonded interactions: 0.50 for Al(s)–O, 0.55 for Al(l)–O, and 0.45 for Al(l)–Al(l). These cut-off values capture the main bonded atoms while excluding weaker, non-bonded interactions in the tails of the distributions. Thus, in the subsequent bond-counting analysis of oxidation, the selected cutoff values for different atomic pairs are applied.

Figure 9 shows the evolution of Al(l)-O newly formed bonds and Al(l)-Al(l) metallic bonds of the Al droplet on the α-Al_2O_3 substrate during the wetting process at (a) 700 K and (b) 1100 K. In both cases, the number of Al(l)-O bonds increases while the number of Al(l)-Al(l) decreases over the duration of 600 ps,

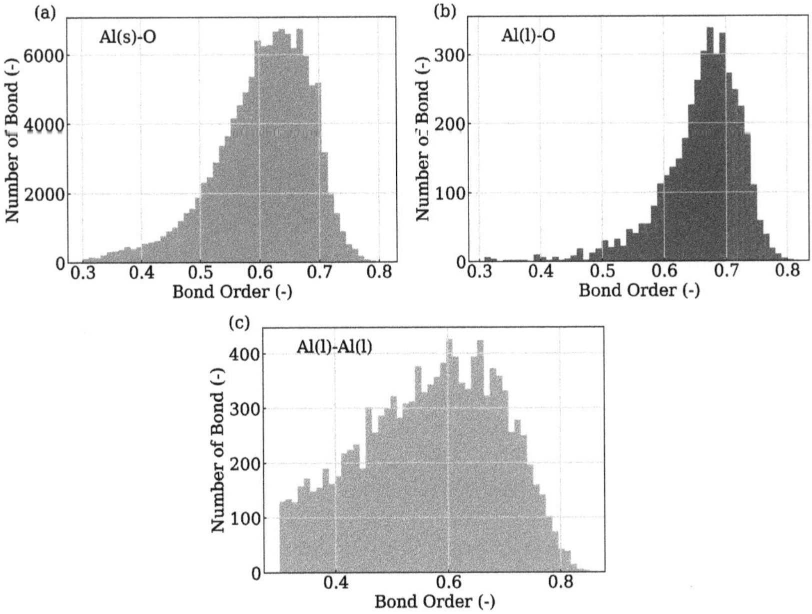

Fig. 8. Bond order distribution for different atom groups at 700 K. (a) Al(s)-O pairs in α-Al_2O_3; (b) Al(l)-O newly formed pairs; and (c) Al(l)-Al(l) metallic bonds in the Al droplet. The time is selected at 175 ps.

highlighting the occurrence of oxidation during the wetting process, in which the metallic bonds are disrupted and new Al(l)-O bonds are formed. Temperature promotes oxidation, as the number of newly formed Al(l)-O bonds at 1100 K is significantly higher than at 700 K. Furthermore, at higher temperatures, the oxidation is rapid and extensive, which significantly reduces the stability of metallic bonds within the droplet. This aligns with the observed temperature-dependent wetting behavior. An interesting feature is that the number of Al(l)–Al(l) bonds reaches a plateau much earlier at 1100 K than at 700 K. This behavior can be understood from the competition between oxidation at the interface and thermally activated restructuring within the liquid droplet. At 1100 K, the higher thermal energy leads to significantly faster atomic diffusion and bond exchange in the liquid Al. The droplet therefore relaxes rapidly towards a quasi–steady coordination state that is compatible with the oxide interface: Al–Al bonds near the interface are replaced by Al–O bonds, while the loss of metallic neighbours is quickly compensated by Al atoms supplied from the interior of the droplet. After this initial adjustment, a dynamic balance between Al(l)–Al(l) and Al(l)–O coordination is established and the total number of Al(l)–Al(l) bonds remains approximately constant. In contrast, at 700 K the lower mobility slows down both diffusion and interfacial restructuring, so that the Al(l)–Al(l) network is

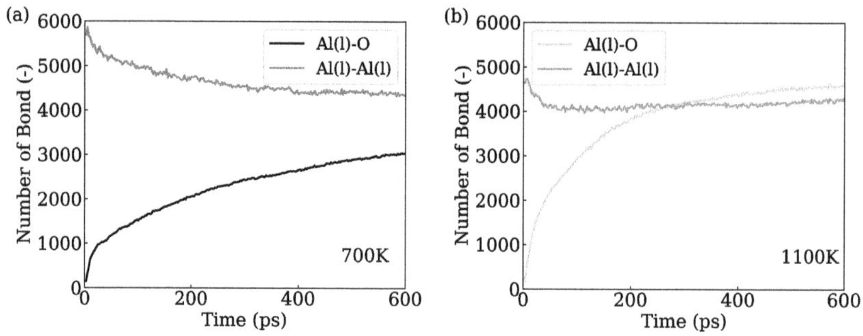

Fig. 9. Variation of number of bond with time at different temperatures. (a) 700 K and (b) 1000 K.

depleted more gradually and the corresponding curve approaches its plateau only at much longer times.

3.3 Effect of Crystallographic Orientation of α-Al_2O_3

Figure 10 illustrates the effect of substrate orientation on the wetting and oxidation behavior of liquid Al droplets on α-Al_2O_3. Four orientations are compared: (a) c-plane (Al-terminated), (b) c-plane (O-terminated), (c) a-plane, and (d) r-plane. The Al droplet initially maintains a nearly spherical shape and progressively spreads with time at all substrates. For the Al-terminated c-plane, as shown in Fig. 10(a), the droplet maintains a more compact shape, indicating weak interaction between the Al droplet and the substrate and thus less oxidation. In contrast, the O-terminated c-plane in Fig. 10(b) demonstrates extensive oxidation at the equilibrium state and lower equilibrium contact angles, which reflect stronger droplet-substrate interactions. Among the orientations, the Al-terminated c-plane shows the least wettable surface, while the O-terminated c-plane, a-plane in Fig. 10(c), and r-plane in Fig. 10(d) all exhibit pronounced spreading and oxidation of the Al droplet. These differences highlight the strong crystallographic dependence of wettability and oxidation. Oxygen-rich or mixed terminations facilitate new Al–O bond formation, whereas Al-terminated surfaces hinder oxidation and maintain higher equilibrium contact angles, as demonstrated in Fig. 10(e).

Figure 11 clarifies the mechanism of the orientation-dependent wetting behavior by the bond analysis. For all substrates, the number of Al(l)-O bonds increases as the wetting process evolves, while the number of Al(l)-Al(l) decreases. However, the rate and extent of these processes depend significantly on the substrate orientation. On the Al-terminated c-plane (in Fig. 11(a)), Al(l)–O bond formation is comparatively slow, while on the O-terminated c-plane, a-plane, and r-plane (Fig. 11(b-d)), Al(l)–O bond formation is significantly accelerated, with the number of Al(l)–O bonds surpassing Al(l)–Al(l) bonds within 200 ps, showing strong chemical reactivity at the interface and within the droplet.

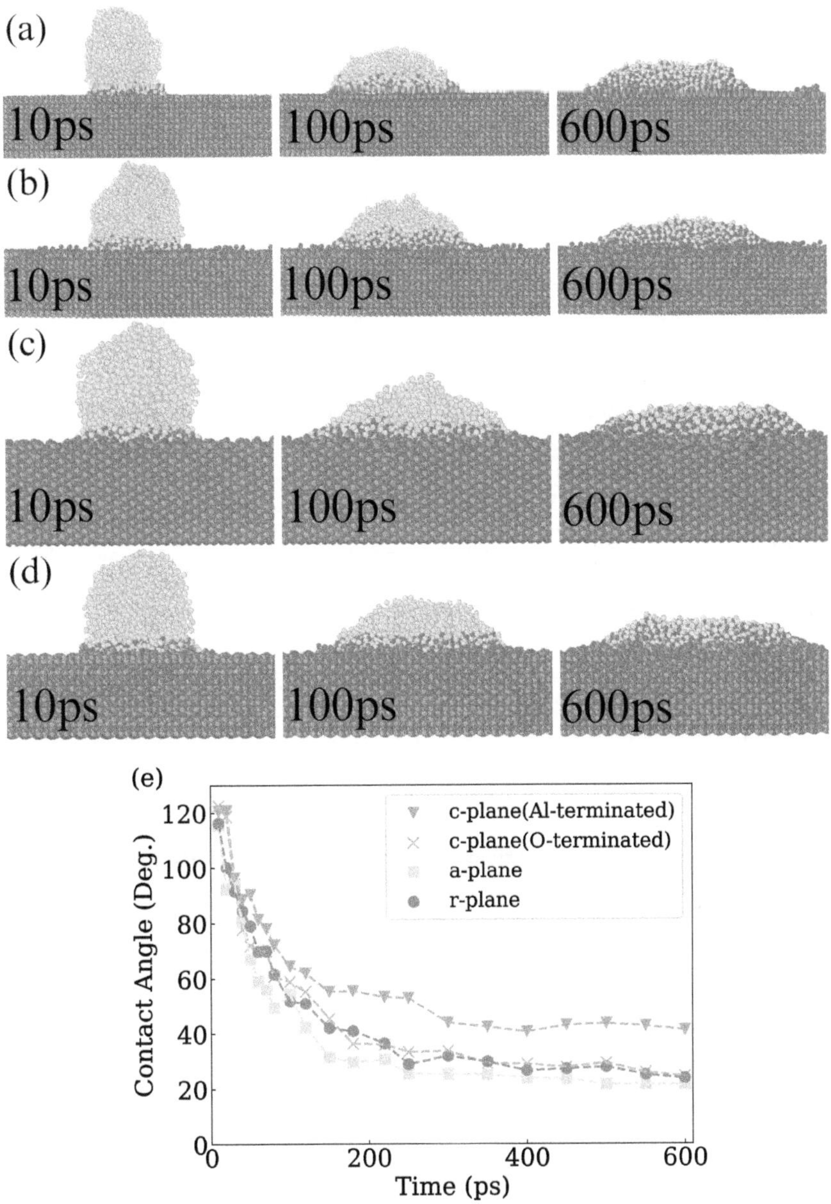

Fig. 10. Illustrations of evolution of Al droplets at the α-Al_2O_3 with different crystallographic orientations. (a) c-plane (0001) with Al-terminated, (b) c-plane (0001) with O-terminated, (c) a-plane (1-100), and (d) r-plane (1-102). (e) Variation of the contact angle for different orientations.

The current simulation results are in good agreement with experimental observations [6], where O-terminated and oxygen-rich surfaces of α-Al_2O_3 have been reported to exhibit higher chemical reactivity with liquid Al, leading to faster spreading and stronger interfacial adhesion. In contrast, Al-terminated surfaces are less reactive, resulting in poorer wettability and weaker bonding.

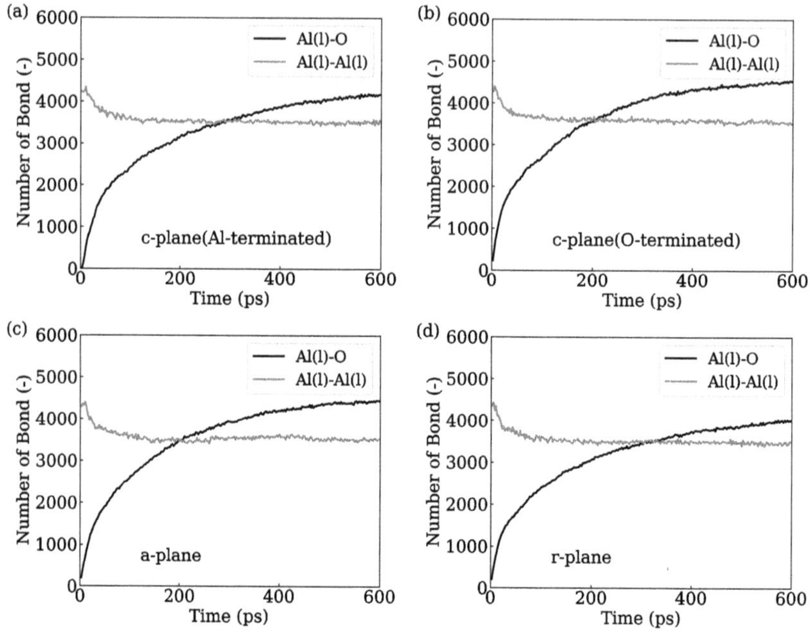

Fig. 11. The effect of substrate orientation on bond number during wetting process.

4 Conclusions

In conclusion, we conducted an atomistic numerical study to study the wetting behavior of Al droplets on α-Al_2O_3 substrates under various conditions. We found that the size of the Al droplet does not influence the wettability of the Al/α-Al_2O_3 system noticeably once it exceeds 18Å. But increasing temperature facilitates the spreading of the droplets and improves wettability as well oxidation. Among the crystallographic orientations examined, Al-terminated c-plane (0001) exhibits the highest contact angle and thus the worst wetting behavior, highlighting the strong substrate orientation-dependent effect. These findings offer a deep insight into the wetting dynamics in the Al/α-Al_2O_3 system, which is important for controlling interfacial properties. We note that the simulated equilibrium contact angles (approximately 20–45°) are significantly lower than those in the experiment. Several factors may contribute to this discrepancy.

First, the simulations are necessarily limited in system size and timescale: the droplet radius is of the order of a few nanometres and the observation time is at most several hundred picoseconds, whereas experiments involve macroscopic droplets that can evolve over much longer times. Line tension and finite-size effects in nanoscale droplets are known to reduce the apparent contact angle compared to the macroscopic value, and the rapid dynamics in our simulations may favour highly wetted metastable states. Second, the oxide structure and chemistry are idealised in the model. We use a perfect α-Al_2O_3 substrate, whereas experimental substrates often contain surface roughness, defects, impurities and non-stoichiometric oxide layers, all of which can effectively reduce the substrate surface energy and increase the measured contact angle. Third, the interaction energetics are determined by the chosen ReaxFF parameterization. Although this force field reproduces a range of bulk and interfacial properties for Al/Al_2O_3, it may overestimate the Al–O adhesion or underestimate the metal/oxide interfacial energy, which would systematically enhance wetting in the simulations. Finally, in the present work we neglect external factors such as ambient atmosphere and possible contamination layers that are often present in experiments and typically act to decrease wetting.

References

1. Sharma, A.K., Bhandari, R., Aherwar, A., Rimašauskienė, R., Pinca-Bretotean, C.: A study of advancement in application opportunities of aluminum metal matrix composites. Mater. Today Proc. **26**, 2419–2424 (2020)
2. Hong, S., van Duin, A.: Molecular dynamics simulations of the oxidation of aluminum nanoparticles using the ReaxFF reactive force field. J. Phys. Chem. C. **119** 17876–17886 (2015)
3. Yousefi, E., Sun, Y., Kunwar, A., Guo, M., Moelans, N., Seveno, D.: Surface tension of aluminum-oxygen system: a molecular dynamics study. Acta Mater. **221**, 117430 (2021)
4. Tahani, M., Postek, E., Sadowski, S.: Diffusion and interdiffusion study at Al- and O-terminated Al2O3/AlSi12 interface using molecular dynamics simulations. Mater. **16**, 4324 (2023)
5. Zhuo, J., et al.: The effect of crystallographic orientation of α-Al2O3 on the wetting behavior and adhesion characteristics of aluminum droplets. J. Phys. Condens. **36**, 195001 (2024)
6. Bao, S., Tang, K., Kvithyld, A., Tangstad, M., Engh, T.A.: Wettability of aluminum on alumina. Met. Mater. Trans. B. **42**, 1358–1366 (2011)
7. Shen, P., Fujii, H., Matsumoto, T., Nogi, K.: Reactive wetting of molten Al on different oriented α-Al2O3 single crystals at high temperatures. Scr. Mater. **49**, 563–569 (2003)
8. Shen, P., Zhang, L., Wang, Y.: Influence of oxidation on contact angle between liquid aluminum and Al2O3. Light Metals **827–832**, 2016 (2016)
9. Xu, P., Sun, S., Sun, S., Tan, X., Ren, Y., Jia, H.: First-principles study of the initial stage of aluminum oxidation. Chem. Phys. Lett. **744**, 137237 (2020)
10. Lyu, S., Merkert, N.: Investigation of an Al/Cu wetting system: a molecular dynamics study. CMSE **16–27**, 2025 (2025)

Recent Trends and Challenges in Simulation Science

Nina Merkert(✉) and Alexander Herzog

Simulation Science Center Clausthal-Göttingen, Clausthal University of Technology, Arnold-Sommerfeld-Straße 6, 38678 Clausthal-Zellerfeld, Germany
Nina.merkert@tu-clausthal.de

Abstract. We analyze recent trends and challenges in the field of modeling and simulation for optimizing networks and materials engineering, two major research fields of the Simulation Science Center Clausthal-Göttingen (SWZ). In the context of networks, we emphasize the importance of simulation as a tool for understanding complex telecommunication systems, traffic logistics, and energy and production networks, where traditional analytical methods often fail. With the paradigm shift brought about by Industry 4.0, advanced simulation-based planning that incorporates event-driven and agent-based models are necessary in order to enhance efficiency and adaptability in production and logistics.

In the realm of computational materials engineering, the paper discusses the interdisciplinary nature of materials science, which relies on a multitude of simulation approaches tailored to various scales and phenomena, from atomic behavior to macro-level material characteristics. We address significant trends such as the application of AI and high-performance computing in materials science. Critical challenges facing the simulation and materials communities are the effective management of large datasets, the complexity of bridging multiple scales, and uncertainty quantification that is a crucial topic in both simulation and AI-driven science. We emphasize the necessity for a cultural shift towards data sharing and improved long-term support for data-driven initiatives. This includes the need for standardized practices in data management and the development of digital twins for predictive analysis in dynamic environments. As the fields of simulation and computational materials science continue to evolve, fostering interdisciplinary collaboration and addressing existing challenges will be key to unlocking their full potential and driving innovation across diverse industries and societal applications.

Keywords: Simulation Science · Network Optimization · Materials Simulation

1 Introduction

In this review paper, we will analyze the recent trends in modeling and simulation for optimization of networks and materials engineering, and their challenges. We will discuss important topics for future research that are currently research focus of the Simulation Science Center Clausthal-Göttingen (SWZ).[1] SWZ is a joint scientific facility

[1] www.simzentrum.de

N. Merkert and A. Herzog (Eds.): SimScience 2025, CCIS 3027, pp. 55–68, 2026.
https://doi.org/10.1007/978-3-032-29720-4_5

of Clausthal University of Technology and Georg-August University Göttingen. The aim of SWZ is to strengthen research in the field of simulation science at both universities and to create greater visibility for the subject area. For this purpose, SWZ, founded in 2013, offers various exchange formats and supports the implementation of joint research projects.

Modeling and simulation of real processes is an ongoing central topic within a steadily increasing number of research areas. Simulation is used, if performing examinations on the real system is too expensive, too dangerous or impossible. Simulations means that the investigations are carried out on a model rather than on the actual system. The model can be a physical model like a smaller version of the real object in a wind tunnel. However, nowadays the term "simulation" is usually equated with computer simulation, see Fig. 1. The generic term "computer simulations" covers various concepts that are used for different issues. While numerical simulations usually are used to model physical processes (e.g. in materials science), logistic and production processes as well as timetables and flight schedules are often modeled using event-driven stochastic simulation. If the modeled objects have their own goals and intentions, e.g. when modeling pedestrian flows, agent-based simulations are employed. If large-scale, possibly cybernetically linked systems, e.g. socio-economic systems, are to be investigated, methods of system dynamic simulation are often used.

New concepts such as Industry 4.0 [Bau23] or stress or capacity tests for critical infrastructures (e.g. railroad stations [Ign2021, Tan24] and emergency departments [Bou23]) manifest the steadily increasing social and industrial relevance of the topic of modeling and simulation. Simulation models to answer specific questions are used in all areas of industry as well as for many societal issues. The process behavior of numerous applications under different conditions and external influences needs to be analyzed in advance and hypotheses need to be proven. Innovative modeling approaches have to be developed and optimized within the framework of interdisciplinary cooperation.

The focus of modeling today is on empirical and model-based approaches on the one hand and data-driven approaches such as methods of artificial intelligence using high-performance computing resources on the other. Empirical and model-based approaches describe the development of a system on the basis of laws, e.g. of physics. Material simulations are thus characterized by a variety of different simulation methods, which are tailored to the respective problem class and the corresponding length and time scales. Predictive methods are based on the analysis of large amounts of data. One of the big challenges in multiscale simulation is the integration of methods from quantum mechanics to continuum mechanics that is necessary to improve material design and processing techniques. Machine learning methods are increasingly used when analytics and conventional modeling reach their limits and have become established for predicting structure-property relationships and for material development. Here, data-driven approaches have the advantage that they are independent of the method used to generate data and thus can help to bridge different scales and combine data from experiments and simulations.

2 Recent Trends in Simulation Science

In the following two sections, we describe the trends and challenges in modeling and simulation for the optimization of networks and in materials engineering.

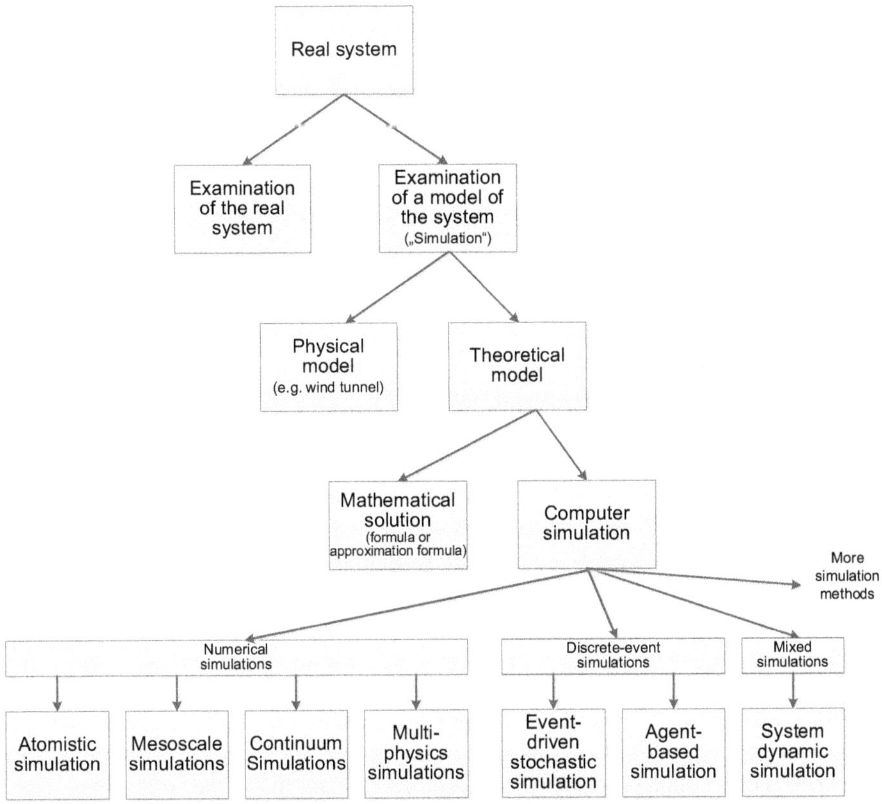

Fig. 1. Overview of simulation strategies

2.1 Simulation and Optimization in Networks

Among the many application areas of simulation science, networked systems represent one of the most mature and industrially relevant fields. Telecommunication networks, traffic systems (e.g. road networks but also flight plans), logistic and production networks, as well as energy networks, share many common characteristics, see Fig. 2.

The complexity of these networks, with their numerous parallel nodes and the dynamic flows between them, is difficult to fully understand and often cannot be accurately predicted or controlled using methods other than simulation. The actual behavior of such networks frequently differs from the originally assumed behavior. The effects of changes in particular are difficult to assess statically. Constructing, operating, modifying, and optimizing these networks usually involve significant infrastructure efforts and associated costs. To avoid undesirable outcomes, simulation is used as a crucial tool to determine the properties of a network and to assess the behavior of critical performance characteristics and parameters at an early stage. With the emergence of Industry 4.0 [Bau23, Bra18, Rei17] concepts such as a significantly more flexible production of small batches [Koe20], right down to a batch size one production, the need for simulation-based planning and optimization of manufacturing and logistics processes is increasing.

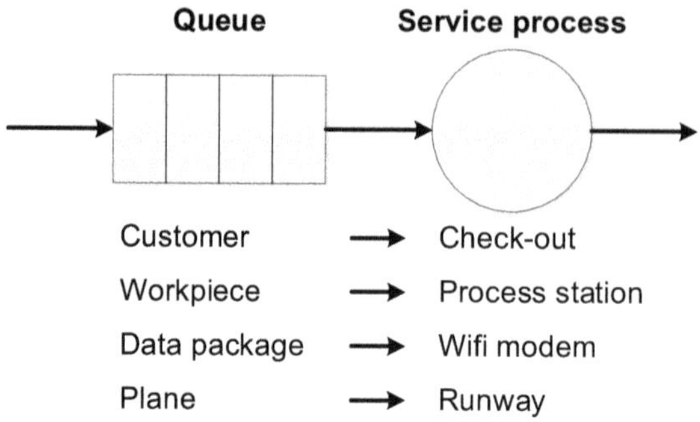

Fig. 2 Applications of discrete-event simulations to real-world problems

This includes the planning of production facilities, including virtual production start-up [Jae24, Mey20], as well as the continuous analysis and optimization of material flows [Erl20], employee deployment [He17] and job sequencing and delivery reliability.

In the industrial environment, event-driven simulation is usually used for this purpose. In this case, the processes are modeled by defining control and routing rules at the stations. The workpieces are passed from one station to the next as passive objects. This corresponds to the pre-industry 4.0 way of thinking. However, Industry 4.0 assumes that the manufacturing intelligence is located in the workpieces themselves. This means that each workpiece carries information about its planned processing steps (such as machining order or target specifications) and can autonomously decide its route through the production system. They tell the stations what work is to be carried out on them and they know their path through production. This approach is familiar in traffic planning. Here, the traffic participants are the intelligent objects that find their own way through the road network on the basis of static specifications and, above all, by considering the current conditions (e.g. traffic jams). In this case, one speaks of an agent-based simulation. In industrial production a mixture of both concepts can be found. While some routing rules are defined by the production network itself (e.g. by the location of the process stations along the conveyor belt) some properties are carries along the production process at the workpieces itself (e.g. the desired color for the paint job).

While classic event-driven simulation is designed for high throughput in terms of the number of simulated objects (and therefore high statistical stability), agent-based simulation enables complex strategies to be mapped. The future challenge is to combine the two concepts if necessary or to select the variant that offers the most advantages for the respective problem.

Another challenge is the coupling of simulation models with optimization. One approach derived from mathematical optimization is to assume the simulation as a function $f(x)$ and to understand the optimization as a higher-level $f(x) = max$ (simulation embedded in optimization). However, other approaches are also conceivable [see VDI3633]: It may be necessary to perform an embedded optimization in a simulation step to determine the respective next actions, or a classical (fast) optimization can be used to restrict the optimal range of values for certain parameters in advance based on a simple model.

The exact optimum for the parameters can then be determined on the basis of a slower simulation based on a more precise model.

Digital twins [Sch24] are being used in advance in planning in an increasing number of areas. The aim of these digital twins is not to create a complete copy of the real system, but rather the twins should merely replicate the behavior of the real systems with regard to certain properties (the model context) in a way that experiments can be carried out on the twin. This is why the terms digital specialist twins or, in the production planning environment, digital logistics twins, are used. In the production environment, the aim is to be able to evaluate the effects of purchasing a new machine etc. or changing the product mix in advance. In traffic planning, both the small-scale (e.g., the traffic light control at an intersection) and large-scale (e.g. new bypass road) effects of changes to traffic routing should be assessed in advance. In urban planning, especially in the context of climate change, it is often a question of being able to evaluate air currents or heat dispersion. What all applications have in common is that a digital twin should help to assess structural changes, which are either expensive or at least difficult to revise retrospectively, in advance without incurring any costs or risks. This evaluation is particularly concerned with identifying dependencies that would not have been recognizable in most cases without a simulation.

In many cases, effects are created on different levels, but they all have an impact on the final result. Each of these levels requires a different modeling and a different simulation. In some cases, event-driven and agent-based simulation have to be used at different levels. Examples include the local processes on a machine versus the entire material flow through a large production process. Or the traffic light control at a single intersection versus the traffic flow in an entire city. The coupling of the findings from the lower, detailed levels with the upper, coarser levels is addressed in the context of the multi-level simulation [Del19]. The challenge here is to simulate only those processes that require a high level of detail and to couple the results directly with a coarser, higher-level model that provides a view of the overall process.

A common characteristic of almost all the real processes under consideration is that the durations for individual sub-steps (arrival times, service times, etc.) are not known exactly. While the historical behavior may be recorded exactly, only probability distributions are usually available for modeling the future behavior for these durations. In order to map the stochastic effects in the model, Monte Carlo simulations are used, in which individual, specific points in time, durations or even branching decisions are mapped on the basis of pseudo-random numbers, which are determined using the underlying distributions. Statistical stability of the results is then achieved according to the law of large numbers via the number of independent repetitions of the entire simulation experiment. All individual runs are subject to the same rules and distributions; only the specific pseudo-random numbers generated on the basis of the distributions vary. Whereas in the past, due to the limited computing power available, the question of the number of repetitions required for statistically stable results was in the foreground, today the focus is more on valid modeling as such [Rab08].

2.2 Computational Materials Engineering

Advances in materials science have always played a crucial role in defining the development stage of a society. Materials science is a key area within the German industrial landscape, providing the foundation for numerous innovations across various sectors of industry.

In this field, a highly interdisciplinary approach is well established, encompassing the closely related areas of "Computational Materials Science", "Computational Physics", and "Computational Chemistry". Simulations have evolved into a dynamic and research-driven scientific discipline within materials science and its neighboring natural sciences. This progress is increasingly recognized and actively supported by industry.

The challenges in materials science are diverse and often involve multiple length and time scales [vdG20] (see Fig. 3). As a result, the field of material simulations is characterized by a wide range of simulation methods, each specifically tailored to address different classes of problems:

On the smallest scale, atomic processes are simulated parameter-free on the basis of natural laws in so-called ab-initio simulations. These quantum mechanics methods usually require high-performance computers, but allow to achieve a variety of quantitative results.

On the macroscopic length scale, materials are treated as continuous media, with their behavior described by material parameters that are either obtained through experiments or derived from more fundamental simulations. Because the simulation techniques at this scale tend to be fairly standardized, commercial software packages are commonly used. The primary focus at this level is on modeling the problem accurately, determining the relevant parameters and equations, and verifying and visualizing the simulation results.

Between the macroscopic and microscopic scales, there are several other simulation methods. These include Monte Carlo methods, molecular mechanics, molecular dynamics, phase field theories, microstructural simulations of dislocation networks and grain boundaries, as well as granular methods. In granular methods, interactions between larger grains are simulated based on Newton's equations of motion.

Therefore, one of the "Grand Challenges in Computational Materials Science" is the multiscale simulation, which ideally extends from ab-initio simulation to the simulation of forming processes and manufacturing processes. Firstly, this challenge is addressed by the fact that simulation parameters are deliberately transmitted from the microscopic simulations to the macroscopic simulations. While first principles simulations only need the material composition, structure and external conditions (pressure and temperature) as input, simulations on the continuum scale require the complete information about the material [vdG20]. Secondly, different simulation methods are integrated into a uniform simulation environment in order to describe effects in which different length and time scales can no longer be decoupled. These activities require the cooperation of different scientific disciplines.

Unlike many other application areas of simulation methods, simulations in materials science often require utilizing all available information processing resources to their fullest extent in order to produce meaningful results. Therefore, questions related to algorithmic complexity and the efficiency of implementations are critically important for advancing simulation techniques in this field.

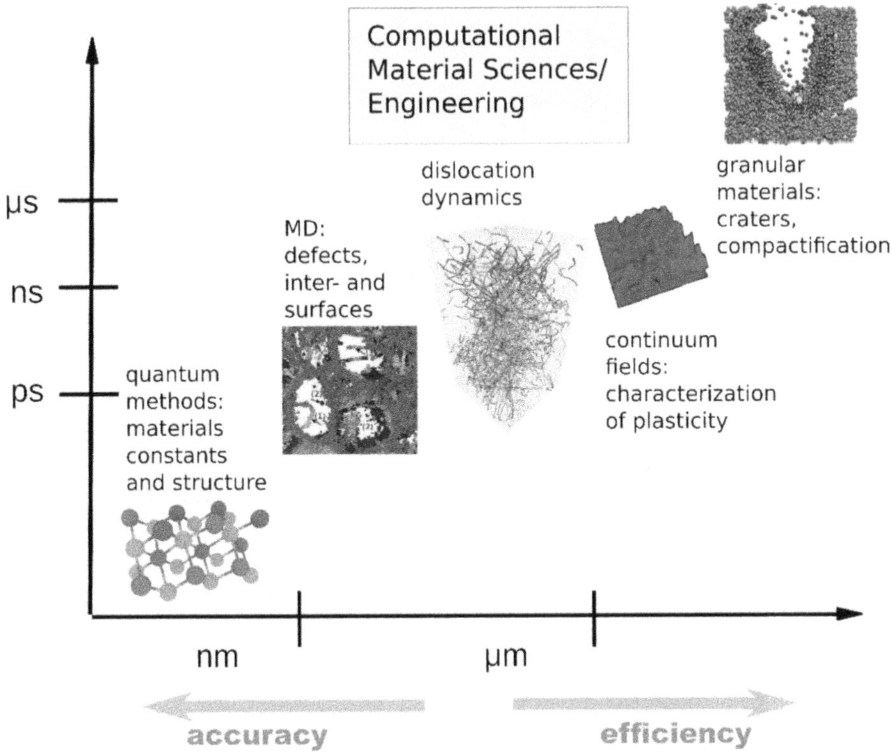

Fig. 3. Length and time scales of multiscale simulation

Because the optimization and characterization of materials involve large volumes of data in various formats, artificial intelligence methods are increasingly employed for data analysis. These methods can generate data when traditional analytics and conventional modeling reach their limits. In recent years, machine learning techniques have become established tools for predicting structure-property relationships [Ech14] and for facilitating material development [Sch19]. However, the accuracy and predictive power of these methods heavily depend on the availability of large datasets, which means that certain aspects of material behavior, e.g. defect dynamics and non-equilibrium processes, are still better described through simulation. In particular, methods like kernel ridge regression require the inversion of a kernel matrix whose dimension equals the number of training samples, so the computational cost and memory usage scale poorly. This makes such methods difficult to apply directly to very large datasets without approximations or specialized numerical techniques. Other models like neural networks can handle larger datasets but are better suitable for input features that exhibit certain correlations, e.g. for image analysis. Another risk is the lack of interpretability. As the exact way input data informs a machine learning model is often determined implicitly by the optimization process and obscured by internal parameters, better methods are needed to interpret and explain the model's decisions [Him19]. A further risk of AI is overfitting which leads to models that appear highly accurate on training data but fail on truly new or different data, giving overly optimistic performance estimates and poor real-world reliability. On the other hand, underfitting produces models that are too simple to capture essential patterns, resulting in systematically biased predictions and missed scientific or practical insights.

3 Recent Advances

3.1 Simulation and Optimization in Networks

In addition to scale bridging (multi-level simulation), the coupling of different simulation methods and the coupling of simulation and data-driven machine learning approaches is a very important current challenge [Geh22, Umo24]. The aim is to couple the speed of AI methods (for mapping large networks) with the accuracy and validity of model-based simulation approaches.

While AI methods based on historical data enable reliable interpolation, only limited forecasts can be made about scenarios that do not exist in the training data (extrapolation) [Fab24]. Model-based approaches, on the other hand, also allow predictions to be made about new, previously unknown scenarios as long as they are still in the model context and the validity of the model as such has been proven before.

Future combined approaches will either decide dynamically when to use machine learning methods and when to use simulation methods, or simulations will be used to generate synthetic input data for AI models.

At the SWZ, the aforementioned questions are currently being addressed in two projects together with industrial partners[2] to answer real questions.

3.2 Data Management in Materials Engineering

In materials science research, very different methods (simulations, microscopic methods and mechanical experiments) are often combined, which also operate on very different size and time scales. As described above, this multiscale character is caused by the highly heterogeneous microstructures present in almost all materials – from crystal defects on the atomic scale to macroscale pores. Since these microstructures are usually not in equilibrium, the interplay of microstructural features on different size and time scales leads to highly path-dependent processing steps. This requires very precise and controlled processing and production, including corresponding documentation [Pra18]. In recent years, in order to develop uniform standards for the storage of research data, many initiatives have been starting. In 2021, the National Research Data Infrastructure for Materials Science and Engineering (NFDI-MatWerk) was established in Germany [Ebe21]. The Fraunhofer Group "Materials, Components (Materials Data Space)" pools the expertise of the 15 materials science-oriented institutes of the Fraunhofer-Gesellschaft. Materials Data Space was launched in 2017 and is the basis for materials development, production and processing within the Industry 4.0 paradigm [Agr16]. In Europe, the initiative NFFA EUROPE sets out a platform to carry out projects for multidisciplinary research at the

[2] https://www.simzentrum.de/en/research/projects/digitaler-zwillinghttps://www.simzentrum.de/en/research/projects/basf

nanoscale extending from synthesis to nanocharacterization to theory and numerical simulation including a data management repository.[3] The Research Data Alliance (RDA) is a world-wide organization with the vision that researchers and innovators openly share data across technologies, disciplines, and countries.[4]

In these consortia, a semantic description and linking of the research data and concepts for cataloging and describing the data are developed. On the one hand, this is accomplished by linking metadata and, on the other, by developing ontologies, for example within the framework of the European Materials Modeling Ontology (EMMO) [Ghe20]. Ontologies in data management are usually formally organized representations of a set of terms and the relationships between them in a specific subject area. They contain inference and integrity rules, i.e. rules for drawing conclusions and ensuring their validity. In contrast to a taxonomy, which only forms a hierarchical structure, an ontology represents a network of information with logical relations.

Although the amount of data produced in each individual measurement or simulation can be enormous, often only a fraction of the data is necessary to arrive at a comprehensive representation of the material properties. Therefore, the data is usually reduced in order to filter out the data that appears to be the most valuable. Big data analysis methods are used for this purpose, which use machine learning methods to predict which data is most important for management.

This is frequently done in order to achieve transparency and efficiency for the development of new materials. Based on the datasets extracted as a basis for further use, the requirements profile of the "hypothetical" new material must be specified from a materials science perspective. With the help of this specification, feature extraction is carried out by means of machine learning methods. Then, systematic material technology tests are defined in order to provide a virtual test environment as part of a digital twin.

The approach of using data science methods for materials sciences is often referred to as the fourth paradigm of materials science in addition to empirical, theoretical and computational science.

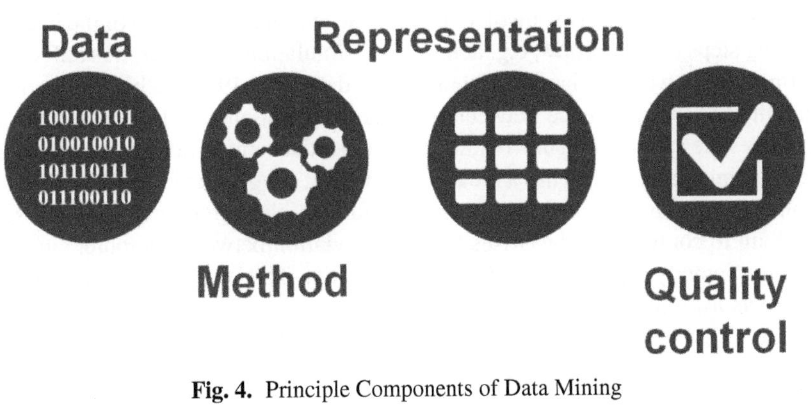

Fig. 4. Principle Components of Data Mining

[3] www.nffa.eu
[4] www.rd-alliance.org

Generally, the basic structure of data-based modeling consists of the following elements: Data, methods, representation and quality control (see Fig. 4). First, one needs to answer the question what data is required. Methods can be e.g. self-learning algorithms, clustering methods, regression and many more. The representation then consists of the unique construction from the data, which must be computationally efficient. Finally, the quality control is a consistency check. For example, we can apply this principle to the D2C (Discrete to Continuous) data science method for transferring discrete dislocation data to the continuous scale [San15]. This conversion method is a concise method for detailed quantitative analysis of dislocation microstructures. It can be used for computationally analyzing discrete dislocation microstructure from various methods, e.g. MD simulations [Gun17], TEM (Transmission Electron Microscopy), continuum or DDD (Discrete Dislocation Dynamics) simulations [San15, Dem23]. Here, the data are initially the positions and Burgers vectors of the dislocations. The method is a coarse-graining method and the representation is continuous fields as the dislocation density. It was used to benchmark the kinematic consistency of continuous dislocation dynamics by Sandfeld et al. [San15]. In the latter paper, the discrete dislocation data from dislocation dynamics simulations was compared to continuous simulations.

3.3 Challenges in Data-Driven Science and Simulation

The 4 Vs, Volume, Velocity, Variety and Veracity play an important role in both data management and the prediction of material simulation data using AI [Him19]. Volume concerns the handling of large data quantities and velocity the time throughput in handling large data streams. Bottlenecks arise in particular with real-time data analysis from simulation outputs. Variety arises from the heterogeneous character of data, which is mainly due to the variety of methods and techniques used and the multiscale character. Veracity, on the other hand, is more difficult to grasp and can mean both the distinction between real and invented data as well as the reproducibility of the data, which can be traced through standardized workflows [Pra18, Him19]. Here, the use of data processing graphs is encouraged. This involves creating an automated protocol of the individual processing steps of a workflow, e.g. a multiscale simulation. However, it is still difficult to distinguish between variance and bias of the data [Him19]. The bias of the results is quantified as the difference between the average result and the truth while the variance defines the spread of results obtained from identical simulations or measurements. Since experimental data also has a measurement error, AI-generated data based on these experiments will also exhibit some variance and bias. Since both experimental setups are difficult to control and bias arises from the deviation between computer simulation and reality, veracity is particularly difficult to quantify. This applies to all methods from experiments to simulation to data-driven approaches.

These 4 Vs refer not only to AI models but to simulation models in general. Because these are also based on data inputs to accurately represent the real-world system. This can be a challenge, especially when dealing with complex systems or when data is limited or unavailable. Inaccurate or incomplete input data can lead to unreliable simulation results. Usually, simulation models in materials science contain a large set of parameters that need to be calibrated based on experimental or theoretical results. This is the case, for example, when complex interaction potentials are fitted for molecular dynamics simulations. The tuning of parameters is a complex and iterative process.

Of course, the computational resources are also a crucial factor especially when dealing with large-scale or highly detailed models. This can limit the analysis and evaluation of both simulation and AI data.

In both simulation and AI-driven materials science, uncertainty is a crucial topic. On the one hand, the simulation models themselves are based on model assumptions that entail uncertainty in comparison to real-world conditions. On the other hand, uncertainties in AI models are due to the lack of complete information and the fact that data or results are subject to randomness and variability. In both cases, systematic errors in the computational methodology are introduced. Since both the parameters for the simulation and the data used for the AI model are generally also based on experiments, experimental uncertainties also play a role. Although sensitivity analysis and uncertainty quantification techniques can help to overcome this limitation, there will always be a certain degree of uncertainty associated with simulation modeling.

The more complex the models become, the more difficult it is to interpret the results and the underlying causes and dynamics of the material behavior can no longer be determined. This applies in particular to AI models, which are not based on physical laws but are often regarded as a kind of black box.

Another challenge is the longevity and maintenance of simulation models and AI developments. Simulation models must be constantly maintained and updated in order to remain usable. The same applies to databases and data management as well as the AI models that are based on the data. Although there are many funding opportunities for data-driven sciences, these are usually only short-term. Long-term financial support for sustainable development is not guaranteed.

Another obstacle is the lack of a distinct culture of data sharing. Often, only the scientific findings and not the data itself are considered relevant for a publication. However, this limits the availability of data and thus the quality of simulation models. In order to achieve a change, clear rules are needed that define how the data can be used.

4 Conclusion

In conclusion, the main challenges in simulation science are the effective management of large datasets, the complexity of bridging multiple scales and methods, and uncertainty quantification in simulation-based and data-driven approaches. An overview of current challenges summarized in this paper can be seen in Fig. 5. It is thus particularly important to foster interdisciplinary cooperation to face these challenges. Only in this way can simulation science develop its full potential and promote industrial innovation.

This review not only provides a comprehensive overview of current methodologies and challenges but also suggests future research directions that could significantly advance the capabilities and applications of simulation science.

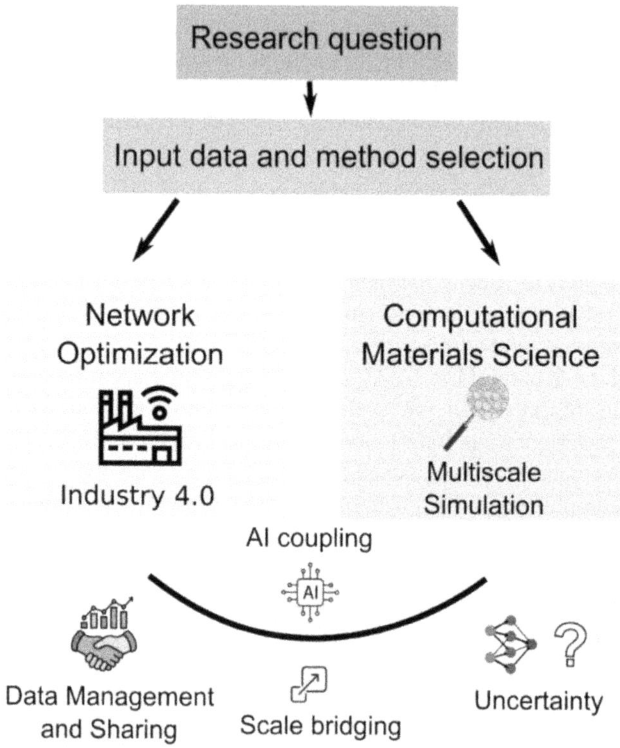

Fig. 5. Overview of current challenges

Literature

Agrawa, A., Choudhary, A.: Perspective: materials informatics and big data: realization of the "fourth paradigm" of science in materials science. APL Mater. **4**, 053208 (2016)

Bauernhansl, T.: Handbuch Industrie 4.0, Band 1: Produktion, Springer, (2023)

Bouramtane, K., Kharraja, S., Riffi, J., El Beqqali, O.: Discrete event simulation for the evaluation of emergency department layout. In: International Conference on Advanced Intelligent Systems for Sustainable Development. AI2SD 2022 Lecture Notes in Networks and Systems, **713**. Springer (2023)

Bracht, U., Geckler, D., Wenzel, S.: Digitale Fabrik, Springer (2018)

Delbrügger, T., Meißner, M., Wirtz, A., et al.: Multi-level simulation concept for multidisciplinary analysis and optimization of production systems. Int. J. Adv. Manuf. Technol. **103**, 3993–4012 (2019)

Demirci, A., Steinberger, D., Stricker, M., Merkert, N., Weygand, D., Sandfeld, S.: Statistical analysis of discrete dislocation dynamics simulations: initial structures, cross-slip and microstructure evolution. Model. Simul. Mater. Sci. Eng. **31**, 075003 (2023)

Eberl, C., et al.: Consortium proposal NFDI-MatWerk. Zenodo. (2021). https://doi.org/10.5281/zenodo.5082837

Echlin, M.P., Lenthe, W.C., Pollock, T.M.: Three-dimensional sampling of material structure for property modeling and design. Integr. Mater. Manuf. Innov. **3**, 21 (2014)

Erlach, K.: Wertstromdesign, Der Weg zur schlanken Fabrik, Springer (2020)
Fabiani, G., et al.: Task-oriented machine learning surrogates for tipping points of agent-based models. Nat. Commun. **15**(1), 4117 (2024)
Gehlot, V., Rokowski, P., Sloane, E.B., Wickramasinghe, N.: Taxonomy, tools, and a framework for combining simulation models with AI/ML models. In: 2022 Annual Modeling and Simulation Conference (ANNSIM), pp. 18–29. IEEE (2022)
Ghedini, E., Goldbeck, G., Friis, J., Hashibon, A., Schmitz, G.: European Materials and Modelling Ontology: Version 1.0.0-beta (2020). https://emmo-repo.github.io/versions/1.0.0-beta/emmo.html
Gunkelmann, N., Alhafez, I. A., Steinberger, D., Urbassek, H. M., Sandfeld, S.: Comput. Mater. Sci. 135, 181 (2017)
Herzog, A.: Callcenter – Analyse und Management, Modellierung und Optimierung mit Warteschlangensystemen, Springer (2017)
Himanen, L., Geurts, A., Foster, A.S., Rinke, P.: Data-driven materials science: status, challenges, and perspectives. Adv. Sci. **6**, 1900808 (2019). https://doi.org/10.1002/advs.201900808
Ignatov, A.N., Naumov, A.V.: On the problem of increasing the Railway Station capacity. Autom. Remote. Control. **82**, 102–114 (2021)
Jaensch, F., Verl, A.: Simulationsmodelle der Virtuellen Inbetriebnahme als Lernumgebung für Reinforcement Learning. In: Verl, A., Röck, S., Scheifele, C. (eds.) Echtzeitsimulation in der Produktionsautomatisierung. Springer Vieweg, Berlin, Heidelberg (2024)
Koether, R., Meier, K.-J.: Lean Production für die variantenreiche Einzelfertigung, Springer (2020)
Meyer, T., Grillitsch, U.: Virtuelle Inbetriebnahme mittels Ablaufsimulation in der Automobilindustrie. In: Mayer, G., Pöge, C., Spieckermann, S., Wenzel, S. (eds.) Ablaufsimulation in der Automobilindustrie. Springer Vieweg, Berlin, Heidelberg (2020)
Prakash, A., Sandfeld, S.: Chances and challenges in fusing data science with materials science. Pract. Metallography **55**(8), 493–514 (2018). https://doi.org/10.3139/147.110539
Rabe, M., Spieckermann, S., Wenzel, S.: Verifikation und Validierung für die Simulation in Produktion und Logistik: Vorgehensmodelle und Techniken. Springer-Verlag, (2008)
Reinheimer, St.: Industrie 4.0, Herausforderungen, Konzepte und Praxisbeispiele, Springer (2017)
Sandfeld, S., Po, G.: Microstructural comparison of the kinematics of discrete and continuum dislocations models. MSMSE **23**, 085003 (2015)
Schmidt, J., Marques, M.R.G., Botti, S., Marques, M.A.L.: Recent advances and applications of machine learning in solid-state materials science. npj Comput. Mater. **5**, 83 (2019). https://doi.org/10.1038/s41524-019-0221-0
Schluse, M.: Experimentierbare Digitale Zwillinge, Konvergenz von Simulation und Realität, Springer (2024)
Tang, J., Yan, J., Sun, Z., Wang, L.: Simulation-based resource allocation for railway container terminals. In: Qin, Y., Jia, L., Yang, J., Diao, L., Yao, D., An, M. (eds). In: Proceedings of the 6th International Conference on Electrical Engineering and Information Technologies for Rail Transportation (EITRT) 2023. Lecture Notes in Electrical Engineering, 1137, Spring (2024)
Umoga, U.J., et al.: Exploring the potential of AI-driven optimization in enhancing network performance and efficiency. Magna Scientia Adv. Res. Rev. **10**(1), 368–378 (2024)
van der Giessen, E., et al.: Roadmap on multiscale materials modeling, modelling Simul. Mater. Sci. Eng. **28**, 043001 (2020)
Wenzel, S.: VDI 3633, Blatt 11, Simulation und Visualisierung. In Symposium Simulationstechnik 2002, (2002)

Application of Self-consistent Field Theory for Modeling Morphologies and Domain Size of Poly(oxyethylene)-Poly(oxybutylene) Diblock Copolymers

Thore Pohl[1] and Michael Fischlschweiger[2](✉)

[1] Chair of Technical Thermodynamics and Energy Efficient Material Treatment, Institute of Energy Process Engineering and Fuel Technologie (IEVB), Clausthal University of Technology, 38678 Clausthal-Zellerfeld, Germany
[2] Institute of Technical Thermodynamics, Karlsruhe Institute of Technology (KIT), Engelbert-Arnold-Straße 4, 76131 Karlsruhe, Germany
michael.fischlschweiger@kit.edu

Abstract. Block copolymers are important in many technical applications due to their ability to undergo microphase separation, resulting in well-defined morphologies. By combining hydrophilic and hydrophobic segments, such as in the poly(oxyethylene)-poly(oxybutylene) system, these materials can be tailored for specific functions like filtration, thermoresponsive behavior, and biocompatibility. Accurate modeling of their morphologies is crucial for designing these materials effectively. Self-consistent field theory (SCFT) is a well-established method for predicting block copolymer morphologies. This study examines ordered morphologies and domain sizes as well as order-order transitions in the poly(oxyethylene)-poly(oxybutylene) system based on varying block composition and polymer chain length. The results are in good agreement with experimental literature data concerning morphologies and clearly highlight the distinct periodicities and properties of the various morphologies.

Keywords: polymer thermodynamics · diblock copolymers · phase behavior · microphase separation

1 Introduction

Block copolymers are versatile because they can form nanostructured self-assemblies through microphase separation [1–5]. A morphology forms by microphase separation when interactions are sufficiently strong [6]. Thus, morphology formation occurs through segregation depending on the segregation strength χN [6]. In the weak segregation limit (WSL), where χN is close to the critical value, the system already exhibits deviations from a homogeneous composition with

N. Merkert and A. Herzog (Eds.): SimScience 2025, CCIS 3027, pp. 69–84, 2026.
https://doi.org/10.1007/978-3-032-29720-4_6

weakly developed periodic morphologies [6,7]. These deviations emerge continuously at the order-disorder transition (ODT), marking a second-order or weakly first-order phase transition [6–8]. By contrast, in the strong segregation limit (SSL), the morphologies are well-defined and domains are strongly segregated with comparably sharp interfaces [6–8]. The equilibrium state of the system is reached through the global minimum of the free energy at a certain temperature and composition [6,7]. These periodic, well-defined morphologies are crucial for many uses [2,3]. Additionally, the chemical adaptability of block copolymers enables customization of material properties by mixing monomers with various chemical properties.

In this context, the block copolymer poly(oxyethylene)-poly(oxybutylene) (POE-POB) is particularly interesting. The block copolymer consists of a hydrophilic POE segment and a hydrophobic POB segment, which confers unique amphiphilic properties that are particularly advantageous for medical technology applications, including tissue engineering and antimicrobial surface coatings for implants [9–13]. For instance, by tailoring the polymer composition, implants can be engineered to meet specific biocompatibility requirements at their implantation site, thereby enhancing their integration and functional performance [10,11,14]. Additionally, the combination of hydrophilic and hydrophobic parts in the POE-POB system is crucial for thermoresponsive materials [15,16]. Temperature-sensitive control allows for precise modulation of the dominant properties between hydrophilicity and hydrophobicity [16]. Moreover, the adjustable solubility characteristics of POE-POB facilitate its application in membrane technology [17,18].

A detailed understanding of the morphologies resulting from microphase separation is crucial for various applications of these block copolymers. This involves examining how morphology formation depends on factors such as block composition, degree of polymerization, and polymer asymmetry. Additionally, describing order-order transitions (OOT) and the domain size of the morphologies is important for determining material functionality. As the domain size is defined as the size of the smallest repeating unit of the periodic morphology and directly influences phase stability and physical properties, understanding this quantity is essential for linking structure and properties in material development.

Computational thermodynamic simulations are vital for understanding these complex relationships and supporting a tailored design for block copolymers like POE-POB. Different thermodynamic equations of state and g^{E}-models have been successfully developed in the past, demonstrating strong capabilities to effectively model phase behavior of polymers [19–21]. They are curcial for polymer design and process development [22–25]. To analyze the micro-level morphologies resulting from nanostructured self-assembly, field theoretical methods are necessary [7,26].

In particular, self-consistent field theory (SCFT) enables the modeling of inhomogeneous concentration fields caused by microphase separation through a mean-field approach to solve the field equations [7]. The polymer chain contour is represented as a variable along the chain axis, and interactions are modeled by effective fields [7]. The effectiveness of SCFT in describing block copolymer morphologies has been shown very successfully in many theoretical studies, especially above the ODT [7,27–29].

To further understand the POE-POB system in the perspective of technological applicability, this work aims to calculate the well-defined morphologies [30] and OOT above the strong segregation limit. Disordered states and order-disorder transitions are not in the focus of this work and are hence not addressed. Herein, estimates of the domain size d will be made, since these data for this system are not yet available according to the author's knowledge. To this end, morphology-specific factors according to Refs. [31–33] will be determined, establishing the relationship between block composition f, degree of polymerization N, and domain size d.

2 Theory and Modeling

First, Subsect. 2.1 presents briefly the theory of SCFT [7,26]. Then, Subsect. 2.2 discusses the calculation of the dimension of the spatial domains based on spatially inhomogeneous concentration profiles using a correlation approach.

2.1 Self-consistent Field Theory

SCFT enables the calculation of spatially inhomogeneous concentration fields [27]. In this way, the morphologies of the block copolymers formed by microphase separation can be calculated [27,29,30]. The SCFT equations are based on the partition function for the respective polymer system [7]. The partition function is defined using the functional integral, which considers one or more chemical potential fields $w(\mathbf{r})$, as shown in Eq. (1) [7,26]:

$$\mathcal{Z} = \int \mathcal{D}w \exp\left(-\mathcal{H}[w]\right) \tag{1}$$

The effective Hamiltonian is expressed as a non-local functional of the field variables and it contains the inverse thermodynamic temperature β, defined as $\beta = 1/(k_\mathrm{B}T)$ [7]. The Hamiltonian $\mathcal{H}[w]$ depends on the interaction and the chain model used, as well as on general polymer properties as for instance the polydispersity and the molecular weight. $\mathcal{D}w$ denotes the integration over all configurations w [7]. Within the SCFT, the Hamiltonian follows the mean-field assumption with Eq. (2) [7]:

$$\left.\frac{\partial\mathcal{H}[w]}{\partial w(\mathbf{r})}\right|_{w=w^*} = 0 \tag{2}$$

The mean-field approximation amounts to the assumption that a single field configuration $w^*(\mathbf{r})$ dominates the integral in Eq. (1) [7]. The application of this approximation transforms the integral from Eq. (1) into numerically solvable equations [7,26]. For the working equations, numerical implementations including pseudo-spectral methods and the Anderson mixing algorithm are provided for instance in the PSCF library in Refs. [27,28] and for their solution in Refs. [34,35]. Convergence is assessed by simultaneously monitoring the change in free energy and the residuals of the self-consistent fields and corresponding segment density profiles [27]. All of these quantities must fall within predefined tolerance thresholds to ensure a stable and accurate self-consistent solution. In addition, for incompressible systems, the vanishing of the internal stress is commonly employed as an additional convergence criterion, confirming equilibrium of the unit cell [36].

In this framework the configuration of the polymer chain in Eqs. (3) and (4) is described by the forward propagator $q(\mathbf{r}, s)$, using the position vector $\mathbf{r}$ and the contour variable s [27,37]:

$$\frac{\partial q}{\partial s} = -H_\alpha q \tag{3}$$

$$H_\alpha f(\mathbf{r}) = -\frac{b_\alpha^2}{6}\nabla^2 f(\mathbf{r}) + w_\alpha(\mathbf{r}) f(\mathbf{r}) \tag{4}$$

In Eq. (4) b_α describes the Kuhn length. The symbols $\alpha, \beta, \ldots, C-1$ describe the different monomer species and ∇^2 represents the Laplace operator applied to the spatial field $f(\mathbf{r})$ [37]. Equations (3) and (4) are denoted as modified diffusion equations (MDE) or Edwards equations [7,27,37]. The following initial condition is applied, given in Eqs. (5) [37]:

$$q(\mathbf{r}, s=0) = 1 \tag{5}$$

The chain segment at the contour origin is assigned an equal probability density at every point in space, representing a spatially uniform initial condition before the imposition of interaction fields or external potentials [7].

The influence of the other polymer chains is considered via the effective field $w_\alpha(\mathbf{r})$ in Eq. (4). This field is defined according to Eq. (6) [27,37]:

$$w_\alpha(\mathbf{r}) = \sum_{\beta=0}^{C-1} \chi_{\alpha\beta}\phi_\beta(\mathbf{r}) + \xi(\mathbf{r}) \qquad \text{with} \qquad \chi_{\alpha\beta} = \chi_{\beta\alpha} \tag{6}$$

Here, $\chi_{\alpha\beta}$ describes the molecular interaction parameter, which is assumed to be symmetrical for the POE-POB system [37]. The Lagrange multiplier $\xi(\mathbf{r})$ is determined numerically under the incompressibility constraint shown in Eq.n (7) [27,28,37]:

$$1 = \sum_{\alpha=0}^{C-1} \phi_\alpha(\mathbf{r}) \tag{7}$$

In addition, the following Eq. (8) applies [27,28]:

$$\phi_\alpha(\mathbf{r}) = v\langle c_\alpha(\mathbf{r})\rangle \tag{8}$$

In Eq. (8), c describes the average concentration $\langle c_\alpha(\mathbf{r})\rangle$ and v is the volume occupied by the monomer. The partition function Q_i of a single polymer chain of species i with length L as a function of the position vector $\mathbf{r}$ and the chain contour variable s is described in Eq. (9) [27,37]:

$$Q_i = \frac{1}{V} \int \mathrm{d}\mathbf{r}\, q(\mathbf{r}, s) q^\dagger(\mathbf{r}, L - s) \tag{9}$$

In Eq. (9), $q^\dagger$ describes the backward propagator and V is the total volume of the system [27]. The invariance of the chain partition function Q with respect to the contour parameter s is ensured numerically through sufficiently fine discretization both along the polymer contour and of the unit cell [27,37]. Convergence is achieved by applying criteria that assess the self-consistent enforcement of the incompressibility constraint and the convergence of the free energy [27,37].

To identify the present morphology, the free energies are also required. The dimensionless free energies $\tilde{f}$ can be calculated according to Eq. (10) [37]:

$$\tilde{f} = \frac{Fv}{Vk_\mathrm{B}T} \tag{10}$$

$$\tilde{f} = \tilde{f}_\mathrm{id} + \tilde{f}_\mathrm{int} \tag{11}$$

F in Eq. (10) represents the Helmholtz energy, and the normalization yields the dimensionless free energy $\tilde{f}$. This is composed of the energy contributions of the ideal chain arrangement $\tilde{f}_\mathrm{id}$ and the interaction components $\tilde{f}_\mathrm{int}$ with other polymers in Eq. (11), which are considered via the effective fields [27,28,37]:

$$\tilde{f}_\mathrm{id} = \sum_{i=0}^{P-1} \frac{\overline{\phi}_i}{L_i} \left[\ln\left(\frac{\overline{\phi}_i}{Q_i}\right) - 1 \right] - \frac{1}{V} \sum_{\alpha=0}^{C-1} \int \mathrm{d}\mathbf{r}\, w_\alpha(\mathbf{r}) \phi_\alpha(\mathbf{r}) \tag{12}$$

$$\tilde{f}_\mathrm{int} = \frac{1}{2V} \sum_{\alpha,\beta=0}^{C-1} \chi_{\alpha\beta} \int \mathrm{d}\mathbf{r}\, \phi_\alpha(\mathbf{r}) \phi_\beta(\mathbf{r}) \tag{13}$$

In Eq. (12), the sum is taken over all polymers P and also over all distinct monomer types C. The expression for $\tilde{f}_\mathrm{int}$ in Eq. (13) contains the molecular interaction parameter $\chi_{\alpha\beta}$ from Eq. (6) [27,28,37].

2.2 Dimension of the Spatial Domains

Experimental investigations show that the morphologies formed by microphase separation exhibit critical domain sizes [2,3]. The unit cell of an ordered system is characterized by a set of parameters θ_i, for example the cell lengths and cell angles [36]. The free energy depends on these parameters. Consequently, at equilibrium the condition for the critical domain size expressed by Eq. (14) must be satisfied [36]:

$$\left.\frac{\partial F}{\partial \theta_i}\right|_{f,\chi N,\ldots} = 0 \tag{14}$$

In order to investigate the inhomogeneous composition in Eq. (15) the local deviation from the average block volume fraction is defined [31,32]:

$$\delta\phi_\alpha(\mathbf{r}) = \phi_\alpha(\mathbf{r}) - \langle\phi_\alpha\rangle \tag{15}$$

The average block volume fraction for the volume V is calculated using Eq. (16):

$$\langle\phi_\alpha\rangle = \frac{1}{V}\sum_{\mathbf{r}}\phi_\alpha(\mathbf{r}) \tag{16}$$

To describe the spatially inhomogeneous composition in Fourier space, the function $g(\mathbf{r})$ in Eq. (17) according to Ref. [32] is defined with respect to the position $\mathbf{r}_0$:

$$g(\mathbf{r}) = \langle\delta\phi_\alpha(\mathbf{r}_0)\cdot\delta\phi_\alpha(\mathbf{r}_0+\mathbf{r})\rangle_{\mathbf{r}_0} \tag{17}$$

Substituting the definition from Eq. (15) in Eq. (17) yields Eq. (18) [31,32,38]:

$$g(\mathbf{r}) = \langle\phi_\alpha(\mathbf{r}_0)\cdot\phi_\alpha(\mathbf{r}_0+\mathbf{r})\rangle - \langle\phi_\alpha\rangle^2 \tag{18}$$

In Fourier space, the structure factor is then defined based on the function $g(\mathbf{r})$ (Eq. (18)) in Eq. (19) [7,28]:

$$S(\mathbf{r}) = \int \mathrm{d}^3\mathbf{r}\ \exp\left(-i\mathbf{q}\cdot\mathbf{r}\right)\ g(\mathbf{r}) \tag{19}$$

This relation includes the correlation function $g(\mathbf{r})$, which can also be calculated from the structural factor function in Eq. (19). This is shown in Eq. (20) [32,37, 38]:

$$g(\mathbf{r}) = \frac{1}{(2\pi)^3}\int \mathrm{d}^3\mathbf{q}\ \exp\left(i\mathbf{q}\cdot\mathbf{r}\right) S(\mathbf{q}) \tag{20}$$

The structure factor $S(\mathbf{r})$ enables the identification of the morphologies. In particular, a set of characteristic reflections in Fourier space, in this case four reflections, is used for cross-checking and confirming the morphological assignment. The domain size d depends on the respective morphology of the material system and thus also on the composition of the blocks f [7,27]. The SCFT strong segregation limit applies [7,29]:

$$d \propto b\cdot N^{\frac{2}{3}}\cdot k_i(f)(\chi N)^{\frac{1}{6}} \tag{21}$$

The function $k_i(f)$ depends on the block volume fraction f and takes the corresponding morphology i into account. The function $k_i(f)$ is not considered in all SSL formulations and is partially included in the proportionality constant [7,29,31,32]. In order to enable a more precise analysis of the effects on the domain size d, $k_i(f)$ is calculated. The different $k_i(f)$ values are summarized in Table 1.

Table 1. Summary of the functions $k_i(f)$ depending on morphology for common periodicities [33].

Periodicity	$k_i(f)$
lamellar	$k_{\text{lam}}(f) = 1$
hexagonal	$k_{\text{hex}}(f) = \frac{1}{\sqrt{f}}$
bcc	$k_{\text{bcc}}(f) = \left(\frac{8\pi}{3f}\right)^{\frac{1}{3}}$
fcc	$k_{\text{fcc}}(f) = \left(\frac{16\pi}{3f}\right)^{\frac{1}{3}}$
gyroid	$k_{\text{gyr}}(f) \approx 1$

In this geometric relationship, $k_i(f)$ relating domain size to the volume fraction f. The lamellar morphology characteristic of symmetric compositions is used as the normalization reference. The introduction of a proportionality constant in Eq. (21) yields Eq. (22) according to Refs. [31,32]:

$$C_{\text{morph}} = \frac{d}{b \cdot N^{\frac{2}{3}} \cdot k_i(f) \cdot (\chi N)^{\frac{1}{6}}} \tag{22}$$

The proportionality constant C_{morph} is denoted morphology factor [29,31,32]. The procedure for calculating the parameters d and C_{morph} based on this is summarized in the flowchart in Fig. 1 [7,31,32].

3 Experimental Data

The composition and morphology data for the calculations are taken from Refs. [9,39–42]. Experimental investigations of POE-POB diblock copolymers with different compositions and degrees of polymerization have shown lamellar (lam), hexagonal (hex), body-centered cubic (bcc), gyroid (gyr), and disordered (dis) morphologies [9,42]. The morphologies depending on the compositions f_{POE} are listed in Table 2.

Table 2. Experimentally identified morphologies selected for different compositions at a temperature of 70 °C [9].

Morphology	bcc	hex	gyr	lam
f_{POE}	0.210	0.286	0.424	0.519

Table 3 summarizes the Kuhn length b and the monomer volume v used in the simulations for POE-POB.

The experimental morphologies were identified using Small-angle X-ray scattering (SAXS) measurements. For this purpose, different block compositions f

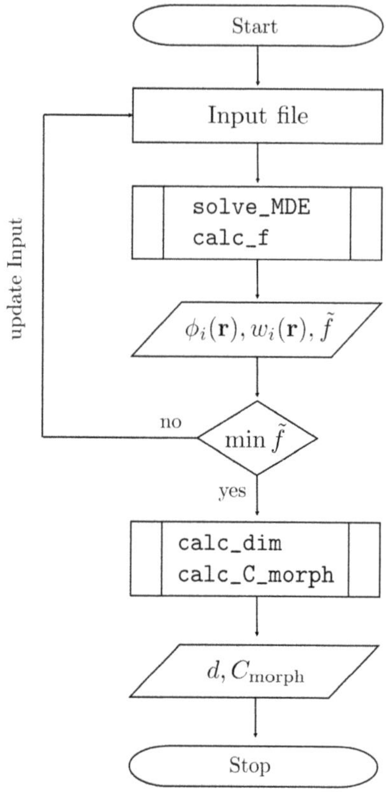

Fig. 1. Flowchart for determining the domain size d and the morphology factor C_{morph} using SCFT and the modeling approach.

of the POE-POB system in Refs. [9,43] were synthesized via anionic polymerization and subsequently measured therein. The measurements from Ref. [9] were performed on the Synchotron Beamline 8.2 of the SRC at the CCLRC Daresbury Laboratory, Warrington, U.K [44]. The composition of the blocks was examined using ^{13}C NMR [9].

When modeling this material system, conformational asymmetry must also be considered [9,45]. For this reason, the asymmetry parameter ϵ is introduced, which considers differences in volume size and also in the stiffness of the blocks [9,43,45]. This is defined by Eq. (23) [9]:

$$\epsilon = \frac{v_{\text{POE}} b_{\text{POB}}^2}{v_{\text{POB}} b_{\text{POE}}^2} \tag{23}$$

For a value of $\epsilon = 1$, conformational symmetry is given. For POE-POB, this results in an asymmetry of $\epsilon = 0.65$ with $\frac{b_{\text{POB}}^2}{b_{\text{POE}}^2} = 1.22$ and $\frac{v_{\text{POB}}}{v_{\text{POE}}} = 1.88$ [9,42].

Table 3. Experimental data as input for the simulation [9,39–42].

i	poly(oxyethylene)	poly(oxybutylene)
b	7,15 Å	7,9 Å
v	0,569 g cm^{-3}	1,067 g cm^{-3}

4 Results

In order to investigate the morphologies formed by the diblock copolymer POE-POB, the morphologies for different block compositions f and χN are first calculated in Subsect. 4.1 considering OOT. Based on these calculations, studies are conducted in Subsect. 4.2 to calculate the domain size d and the morphology factor C_{morph} for different morphologies.

4.1 Calculation of Morphologies

Certain morphologies are calculated and experimentally validated based on Refs. [9,43]. The present morphology is determined by minimizing the free energy. Calculations were performed for the certain temperature and composition for $\chi N = 20$, $\chi N = 30$, and $\chi N = 40$. For $\chi N = 30$ and $\chi N = 40$, the SSL is satisfied. For $\chi N = 20$, this depends on the compositional asymmetry. The discretization was performed in accordance with Ref. [27].

Only calculations above the ODT estimable via Eq. (24) are performed [7]:

$$(\chi N)_{\mathrm{t}} = 10.495 + 41.022\, N^{-\frac{1}{3}} \tag{24}$$

The morphologies calculated in Table 4 agree well with the morphologies reported in Ref. [9]. This clearly indicates that the phase behavior is asymmetric, which is attributed to the inherent asymmetry of the POE and POB blocks [9,42].

OOTs occur both as a result of variations in the block composition and changes in the degree of polymerization [7,30]. For instance, as shown in Table 4, at $\chi N = 20$ the OOT between block compositions $f_{\mathrm{POE}} = 0.30$ and $f_{\mathrm{POE}} = 0.40$ corresponds to a transition from a hexagonal to a gyroid morphology. Likewise, for $f_{\mathrm{POE}} = 0.40$, an OOT is predicted between $\chi N = 30$ and $\chi N = 40$.

For an asymmetric block volume fraction of $f_{\mathrm{POE}} = 0.30$ and $\chi N = 30$, a hexagonal structure is predicted, as shown in Fig. 2.

The cylindrical domains, formed by the minority POE block, arrange into a hexagonal lattice within the continuous majority POB matrix [42]. This two-dimensional structure exhibits pronounced anisotropic properties. The resulting anisotropic channels render these morphologies highly relevant for various information technical applications [2]. Additionally, such morphologies can serve as templates for fabricating periodic nanostructures [2].

Figure 3 shows the calculated gyroid morphology for a block volume fraction of $f_{\mathrm{POE}} = 0.4$.

Table 4. Calculation of morphologies as a function of block composition f and degree of polymerization N, assuming a constant temperature, whereby the experimentally determined interaction parameter $\chi = 0.0864$ is taken from Ref. [9].

f_{PEO}	Structures		
	$\chi N = 20$	$\chi N = 30$	$\chi N = 40$
0.20	dis	bcc	bcc
0.30	hex	hex	hex
0.40	gyr	gyr	hex
0.50	lam	lam	lam
0.55	lam	lam	lam
0.60	lam	lam	lam
0.70	hex	hex	lam
0.80	bcc	hex	hex

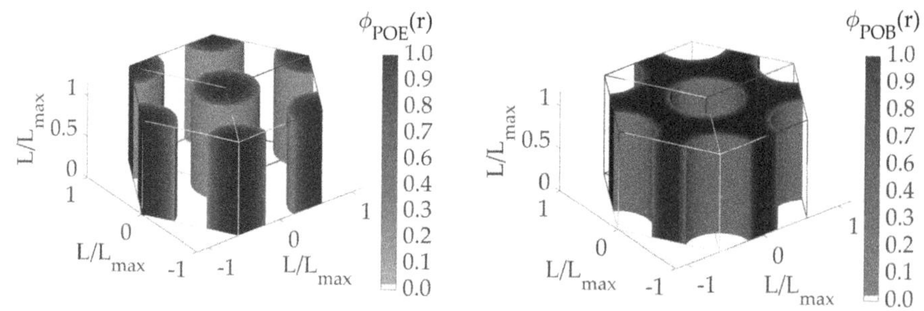

Fig. 2. Hexagonal morphologies of the POE- (left) and the POB-blocks (right) at a block composition of $f_{POE} = 0.30$ and $\chi N = 30$. Three unit cells form a hexagonal cell.

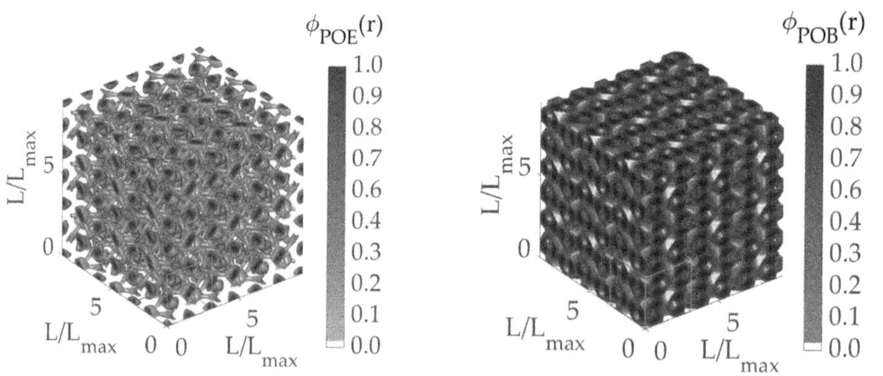

Fig. 3. Gyroid morphologies of the POE- (left) and the POB-blocks (right) at a block composition of $f_{POE} = 0.40$ and $\chi N = 30$. Three unit cells are displayed along each spatial direction.

In this gyroid, both polymer blocks form spatially continuous but inhomogeneously distributed domains, resulting in a large interfacial area. This high interfacial surface is critical for surface functionalization. Gyroid morphologies are particularly relevant in tissue engineering, where they serve as tunable scaffolds [11,13]. Selective removal of one block enables the fabrication of highly porous materials, which is advantageous for applications requiring enhanced permeability and surface area [13].

For block compositions around $f_{\mathrm{POE}} = 0.5$, lamellar structures are formed. These morphologies occur over a broad range of block copolymer compositions due to microphase separation [3,9,42]. The calculated results are presented in Fig. 4. The variation in layer thickness reflects differences in segmental lengths, with the POB layers being thicker than the POE layers [9,39].

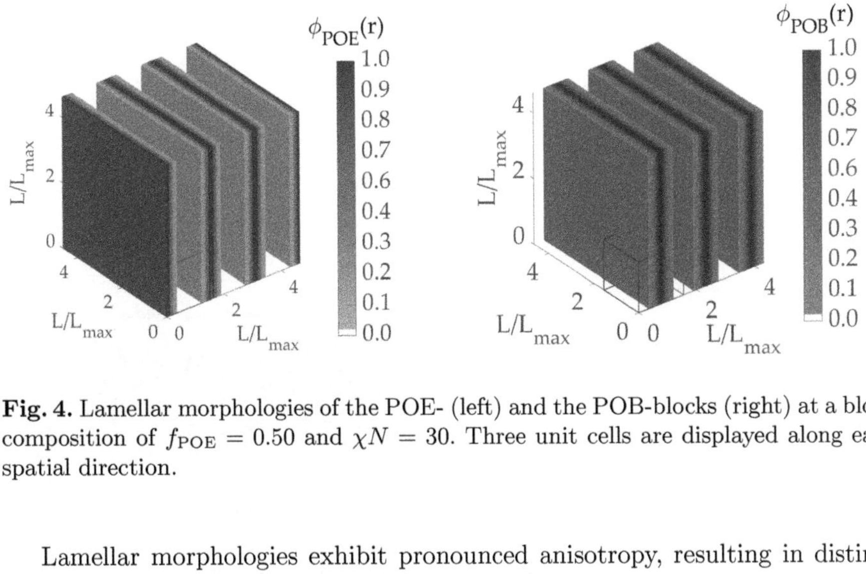

Fig. 4. Lamellar morphologies of the POE- (left) and the POB-blocks (right) at a block composition of $f_{\mathrm{POE}} = 0.50$ and $\chi N = 30$. Three unit cells are displayed along each spatial direction.

Lamellar morphologies exhibit pronounced anisotropy, resulting in distinct mechanical, optical, and electrical properties along directions perpendicular and parallel to the lamellae. Lamellar morphologies are relevant for membrane technology, where the alternating layers function as selective channels [2,17,18]. Furthermore, lamellar morphologies serve as functional interlayers in organic semiconductor devices [2,3].

With a highly asymmetric block composition f, the bcc structure tends to form due to microphase separation [30,42]. This behavior is illustrated for the block composition $f_{\mathrm{POE}} = 0.80$ and $\chi N = 30$ in Fig. 5.

The bcc structure exhibits a high degree of symmetry, with eight neighboring cells surrounding the central domain. The high degree of symmetry makes bcc morphologies particularly attractive for photonics-related applications [2,3].

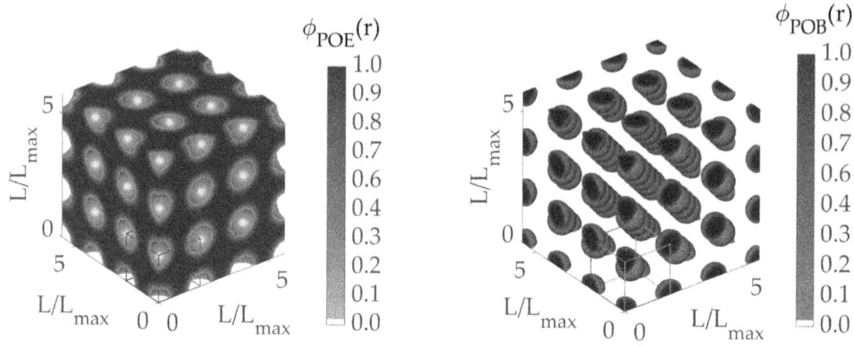

Fig. 5. Bcc morphologies of the POE-(left) and the POB-blocks (right) at a block composition of $f_{\text{POE}} = 0.80$ and $\chi N = 30$. Three unit cells are displayed along each spatial direction.

4.2 Calculation of Morphology Factors and Domain Sizes

Based on the morphology calculations from Subsect. 4.1, the domain sizes of the individual morphologies are examined. The computed SAXS spectra for $f_{\text{POE}} = 0.80$, which shows a bcc morphology, is shown in Fig. 6.

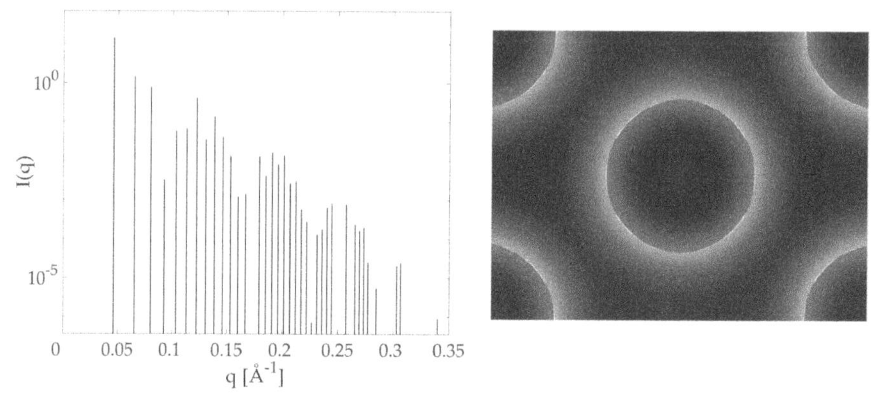

Fig. 6. Modeled SAXS spectrum (left) of the bcc structure at a block composition of $f_{\text{POE}} = 0.80$ for POE-POB, alongside the cross-section (right) along a symmetry axis (red = POB domain, blue = POE domain).

The morphologies and their domain sizes d for different compositions f resulting from the SCFT simulations are summarized in Table 5. Based on these domain sizes d, the morphology factors C_{morph} were determined using Eq. (22).

As shown in Table 5, the morphology factor C_{morph} enables a more accurate comparison of the morphologies according to Eq. (22) than the domain size alone.

Table 5. Modeled domain sizes d and morphology factors C_{morph} for PEO-POB depending on block compositions.

f_{PEO}	$\chi N = 30$	d/nm	C_{morph}/-
0.30	hex	22.45	0.184
0.40	gyr	35.24	0.531
0.50	lam	50.63	0.772
0.80	bcc	13.66	0.098

Overall, it is evident from Table 5 that lamellar structures exhibit the largest spatial periodicity and domain size. In contrast, the bcc structure represents the most compact spatial arrangement for POE-POB. Thus, as compositional symmetry rises, domain size increases according to Table 5. This behavior is in principle alignment with other domain size investigations in the literature carried out for other material systems [33,46].

In the mean-field approximation, the correlation length is infinite, which in practical terms implies that order is maintained throughout the entire system without accounting for defects or interruptions. This assumption is justified from the perspective that the analyzed cases exhibit well-defined morphologies far from the segregation limit in terms of temperature. Also because of $\bar{N} \rightarrow 10^4$ fluctuations are not considered herein [7,8,27].

5 Conclusion

Due to their complex nanostructured self-assembly via microphase separation, block copolymers are highly relevant across numerous application fields [1–5,10–13,18]. This imposes demands on materials engineering to achieve well-defined morphologies. One challenge lies in understanding the domain sizes of the morphologies.

The POE-POB system, composed of hydrophilic POE and hydrophobic POB blocks was experimentally investigated in Ref. [9] and exhibits complex morphological behavior, which makes it hence interesting for different applications [15,42,43]. Morphologies in combination with respective domain sizes have not yet been deeper investigated for this particular material system. Hence, in this work, SCFT-based calculations were performed for the POE-POB system to predict morphologies and domain sizes of block copolymers in the strong segregation limit, incorporating Kuhn lengths and monomer reference volumes as material-specific input parameters [9,39–42].

The predicted morphologies body-centered cubic, lamellar, hexagonal, and gyroid agree quantitatively well with literature data [9,42]. The lamellar morphologies exhibit the largest domain sizes in the POE-POB system, whereas the bcc structures show the smallest domain dimensions. Overall, an increase in domain size is observed with increasing compositional symmetry. This is consistent with other theoretical predictions in the literature [33].

This provides a base for different applications of these morphologies across a broad design space spanning block compositions of POE-POB from $f = 0.20$ to $f = 0.80$ at constant temperature.

In future studies, morphologies should be analyzed at various temperatures. In order to address the spatially limited formation of these morphologies as the temperature increases and approaches the segregation limit, future work could introduce perturbation fields aimed at correcting the mean-field solution.

References

1. Zeiner, T., Fischlschweiger, M.: Diffusion and transport through nanoscale polymer-based coatings. In: Polymer-Based Nanoscale Materials for Surface Coatings, 2023, pp. 291–321 (2023)
2. Yang, G.G., et al.: Block copolymer nanopatterning for nonsemiconductor device applications. ACS Appl. Materials Interfaces **14**(10), 12011–12037 (2022)
3. Yang, C., et al.: Polymer nanocomposites for energy storage, energy saving, and anticorrosion. J. Materials Chem. A **3**(29), 14929–14941 (2015)
4. Xu, Z., Lin, J., Zhang, Q., Wang, L., Tian, X.: Theoretical simulations of nanostructures self-assembled from copolymer systems. Polym. Chem. **7**(23), 3783–3811 (2016)
5. Bates, F.S., Fredrickson, G.H.: Block copolymers-designer soft materials. Phys. Today **52**(2), 32–38 (1999)
6. Leibler, L.: Theory of microphase separation in block copolymers. Macromolecules **13**, 1602–1617 (1980)
7. Fredrickson, G.H.: The equilibrium theory of inhomogeneous polymers, International Series of Monographs on Physics, Oxford Science Publications, Bd. 134, (2006)
8. Fredrickson, G.H., Helfand, E.: Fluctuation effects in the theory of microphase separation in block copolymers. J. Chem. Phys. **87**(1), 697–705 (1987)
9. Mai, S.-M., et al.: Macromolecules **31**, 8110 (1998)
10. Li, P., et al.: Preparation, evaluation and functionalization of biomimetic block copolymer coatings for potential applications in cardiovascular implants. Appl. Surf. Sci. **502**, 144085 (2020)
11. Kutikov, A.B., Song, J.: Biodegradable PEG-based amphiphilic block copolymers for tissue engineering applications. ACS Biomaterials Sci. Eng. **1**(7), 463–480 (2015)
12. Voo, Z.X., Khan, M., Narayanan, K., Seah, D., Hedrick, J.L., Yang, Y.Y.: Antimicrobial/antifouling polycarbonate coatings: role of block copolymer architecture. Macromolecules **48**(4), 1055–1064 (2015)
13. Kim, T.H., et al.: Flexible biomimetic block copolymer composite for temperature and long-wave infrared sensing. Sci. Adv. **9**(6), eade0423 (2023)
14. Smith, J.R., Lamprou, D.A.: Polymer coatings for biomedical applications: a review. Trans. IMF **92**(1), 9–19 (2014)
15. Constantinou, A.P., Wang, L., Wang, S., Georgiou, T.K.: Thermoresponsive block copolymers of increasing architecture complexity: a review on structure-property relationships. Polym. Chem. **14**(3), 223–247 (2023)
16. Liu, R., Fraylich, M., Saunders, B.R.: Thermoresponsive copolymers: from fundamental studies to applications. Colloid Polym. Sci. **287**(6), 627–643 (2009)

17. Signorini, V., Taleb, O., Baschetti, M.G., Minelli, M., Hallinan, D.T.: Effect of block copolymer membrane nanostructure on ethanol-water transport measured by infrared spectroscopy. Polymer **333**, 128637 (2025)
18. Minelli, M., Baschetti, M.G., Hallinan, D.T., Balsara, N.P.: Study of gas permeabilities through polystyrene-block-poly(ethylene oxide) copolymers. J. Membr. Sci. **432**, 83–89 (2013)
19. Gross, J., Sadowski, G.: Application of the perturbed-Chain SAFT equation of state to associating systems. Indust. Eng. Chem. Res. **41**(22), 5510–5515 (2002)
20. Langenbach, K., Browarzik, D., Sailer, J., Enders, S.: New formulation of the lattice cluster theory equation of state for multi-component systems. Fluid Phase Equilib. **362**, 196–212 (2014)
21. Fischlschweiger, M., Enders, S.: A theory for solubility of semicrystalline and branched polymers in one solvent. Macromolecules **47**(21), 7625–7636 (2014)
22. Fan, Z., Schwedes, M., Schwaderer, J., Beuermann, S., Fischlschweiger, M.: Molecular weight as a key for electroactive phase formation in poly(vinylidene fluoride). Materials Res. Lett. **10**(4), 271–277 (2022)
23. Fischlschweiger, M., Enders, S.: Solid-liquid equilibria of crystalline and semicrystalline monodisperse polymers, taking into account the molecular architecture by application of the lattice cluster theory. Mol. Phys. **112**(24), 3109–3119 (2014)
24. Fischlschweiger, M., Enders, S.: Solid-liquid phase equilibria of binary hydrocarbon mixtures predicted by Lattice Cluster Theory. J. Mol. Liq. **212**, 436–443 (2015)
25. Fischlschweiger, M., Danzer, A., Enders, S.: Predicting gas solubility in semicrystalline polymer solvent systems by consistent coupling of Sanchez-Lacombe EOS with a continuum mechanics approach. Fluid Phase Equilib. **506**, 112379 (2020)
26. Fredrickson, G.H., Delaney, K.T.: Field-Theoretic Simulations in Soft Matter and Quantum Fluids, International Series of Monographs on Physics, Oxford Science Publications, Bd. 173 (2023)
27. Arora, A., et al.: Broadly accessible self-consistent field theory for block polymer materials discovery. Macromolecules **49**(13), 4675–4690 (2016)
28. Xie, O., Olsen, B.D.: A self-consistent field theory formalism for sequence-defined polymers. Macromolecules **55**(15), 6516–6524 (2022)
29. Matsen, M.W., Schick, M.: Stable and unstable phases of a linear multiblock copolymer melt. Macromolecules **27**(24), 7157–7163 (1994)
30. Pohl, T.: Makromolekulare Modellierung der Phasenübergänge in Polyoxyethylen-Polyoxybutylen Diblockcopolymeren mit Statistischer Feldtheorie unter Berücksichtigung der Mikrophasenseparation, Master's thesis, Clausthal University of Technology, Institute of Energy Process Engineering and Fuel Technology (IEVB) (2024)
31. Semenov, A.N.: Contribution to the theory of microphase layering in block-copolymer melts. Zh. Eksp. Teor. Fiz. **88**, 1242–1256 (1985)
32. Semenov, A.N.: Dynamical correlation function of polymer density fluctuations in concentrated solutions. XXPhys. A **166**(2), 263–287 (1990)
33. Matsen, M.W., Bates, F.S.: Block copolymer microstructures in the intermediate-segregation regime. J. Chem. Phys. **106**(6), 2436–2448 (1997)
34. Rasmussen, K., Kalosakas, G.: Improved numerical algorithm for exploring block copolymer mesophases. J. Polym. Sci., Part B: Polym. Phys. **40**, 1777–1783 (2002)
35. Rajan, A., Qin, J., Morse, D.: Linear response and stability of ordered phases of block copolymer melts. Macromolecules **41**, 942–954 (2008)
36. Tyler, C.A., Morse, D.C.: Stress in self-consistent field theory. Macromolecules **36**(21), 8184–8188 (2003)

37. Morse, D., Cheong, G.K., Chawla, A., Collanton, R.: Appendix: Modified Diffusion Equations, Online available at: https://dmorse.github.io/pscfpp-man/scft_propagator_page.html. Accessed 15 Aug 2025
38. J.-P. Hansen, I. R. McDonald, Theory of Simple Liquids: with Applications to Soft Matter, Academic Press, 2013, ISBN 978-0-12-387033-9
39. Zhou, Y., Saidi, W.A., Fichthorn, K.A.: Comparison of the binding of polyvinylpyrrolidone and polyethylene oxide to ag surfaces: elements of a successful structure-directing agent. J. Phys. Chem. C **117**(21), 11444–11448 (2013)
40. Zhou, Y., Saidi, W.A., Fichthorn, K.A.: Comparison of the Binding of Polyvinylpyrrolidone and Polyethylene Oxide to Ag Surfaces: Elements of a Successful Structure-Directing Agent, The Journal of Physical Chemistry C (2013)
41. Gerstl, C., et al.: Chain conformation of poly(alkylene oxide)s studied by small-angle neutron scattering. Macromolecules **44**(15), 6077–6084 (2011)
42. Yang, Y.-W., et al.: Structures of oxyethylene/oxybutylene diblock copolymers in their solid and liquid states. Macromolecules **28**(18), 6029–6041 (1995)
43. Mai, S.-M., Booth, C., Nace, V.M.: Specific volumes of poly(oxybutylene) and poly(oxyethylene) dimethyl ethers in the liquid state. Eur. Polymer J. **33**(7), 991–996 (1997)
44. Bras, W., et al.: Simultaneous time resolved SAXS and WAXS experiments using synchrotron radiation. Nucl. Instrum. Methods Phys. Res., Sect. A **326**(3), 587–591 (1993)
45. Matsen, M.W., Schick, M.: Microphases of a diblock copolymer with conformational asymmetry. Macromolecules **27**(14), 4014–4015 (1994)
46. Seul, M., Andelman, D.: Domain shapes and patterns: the phenomenology of modulated phases. Sci., Am. Assoc. Adv. Sci. **267**, 476–483 (1995)

Experimental and Numerical Analysis of Flow-Front Development in Biaxial Non-crimp Fabrics with Temporary Flow Channels

Xiaohui Zhang[1(✉)], Juzhong Zhang[1], Gerhard Ziegmann[1], Purna Jayant Kumar[2], Matthias Overberg[3], and Chokri Cherif[3]

[1] Institute for Polymer Materials and Plastics Engineering, Clausthal University of Technology, Clausthal-Zellerfeld, Germany
xiaohui.zhang@tu-clausthal.de

[2] Institute of Mechanical Engineering, Clausthal University of Technology, Clausthal-Zellerfeld, Germany

[3] Institute of Textile Machinery and High Performance Material Technology, Dresden University of Technology, Dresden, Germany

Abstract. Efficient resin infiltration in complex textile reinforcements remains a persistent challenge in liquid composite molding. This study investigates flow-front development in biaxial non-crimp fabrics incorporating temporary flow channels (TFCs). The proposed approach exploits the fabric architecture and associated dual-scale flow phenomena to improve in-plane permeability. Flow behavior is examined through experiments conducted using a custom in-house setup and numerical simulations employing a dual-domain porous/free-flow model with Volume of Fluid tracking in ANSYS Fluent. Two physics-based modeling simplifications are considered: a representative volume element with periodic boundary conditions and a simplified three-dimensional plate model incorporating User-Defined Function–based porosity assignment. Numerical results show good agreement with experimental observations, demonstrating that TFC-modified fabrics promote flow-front advancement and significantly reduce mold filling time relative to the reference configuration.

Keywords: Liquid composite molding · Biaxial non-crimp fabrics · Numerical simulation · Volume of Fluid · User-Defined Function

1 Introduction

Liquid composite molding (LCM) techniques are widely adopted for producing high-performance fiber-reinforced polymer composites due to reduced material waste, high design flexibility, and scalability for industrial applications [1–3]. Despite these advantages, achieving complete and uniform resin infiltration remains challenging, particularly for thick and multilayer reinforcements [4,5].

N. Merkert and A. Herzog (Eds.): SimScience 2025, CCIS 3027, pp. 85–102, 2026.
https://doi.org/10.1007/978-3-032-29720-4_7

Without auxiliary flow media, incomplete impregnation may cause defects such as dry spots when the resin gels or cures before fully saturating the preform [6].

Among advanced textile reinforcements, biaxial non-crimp fabrics (BX-NCFs) are attractive due to low crimp, high fiber alignment, and enhanced in-plane properties [7,8]. BX-NCFs exhibit dual-scale flow behavior, with resin flowing simultaneously through inter-bundle macro-channels and intra-bundle micro-channels. This dual-scale nature becomes more pronounced in multilayer configurations, where permeability contrasts between scales may lead to non-uniform flow-front development [9,10].

Motivated by the dual-scale effect, integrating temporary flow channels (TFCs) is a promising approach for improving flow control in multilayer textiles [11]. In this concept, selected rovings are locally compacted at regular intervals using sacrificial, resin-soluble threads. The local compaction creates transient preferential pathways that accelerate in-plane flow at early filling stages by promoting transport through inter-bundle regions. As impregnation progresses, the threads dissolve and the textile relaxes toward its original architecture, avoiding permanent voids or channels in the final composite [12]. Previous studies have demonstrated that engineered flow channels and distribution media can significantly improve resin distribution and reduce impregnation times in simplified textile preforms, however, the effectiveness of TFCs has not been systematically quantified [13].

Experimentally, textile effects on flow are often evaluated using global flow-front measurements in one-dimensional (unidirectional) and two-dimensional (radial) tests [14,15]. Numerically, unsaturated flow has been studied using various approaches, yet many works rely on simplified representations and limited validation for complex textile architectures [16,17]. By combining experiments and simulations, this work clarifies the role of TFC-based flow enhancement in BX-NCFs and evaluates a practical modeling strategy to predict flow-front development.

The paper is organized as follows. Section 2 introduces the TFC-integrated BX-NCFs and the experimental characterization. Section 3 presents the multiphase modeling framework and the physics-based simplifications. Section 4 compares experimental and numerical flow-front evolution for reference and modified configurations. Finally, Sect. 5 summarizes the key findings and implications for optimized textile design in LCM.

2 TFC-Integrated BX-NCFs and Characterization Techniques

This section describes the textile concept and the experimental setup used to compare the flow behavior of the reference and TFC-modified architectures.

2.1 TFC-Integrated BX-NCFs

TFC-integrated BX-NCFs provide an effective strategy to increase manufacturing efficiency in fiber-reinforced plastics (FRPs) [11]. The fabric, developed

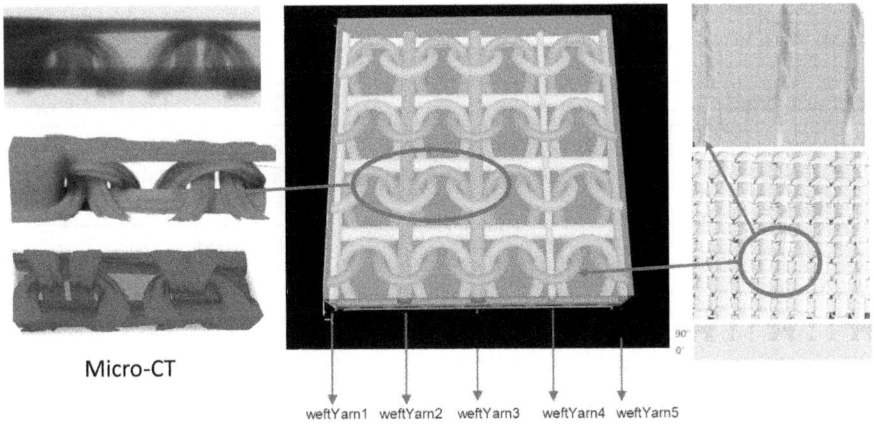

Fig. 1. Yarn configuration and internal architecture of TFC-integrated BX-NCFs.

in collaboration with an industrial partner, was digitally reconstructed from micro-computed tomography (μCT) data and modeled in TexGen to represent the internal architecture [18,19]. Figure 1 illustrates the BX-NCF configuration comprising yarns *weftYarn1* to *weftYarn5*, forming the layered and interconnected structure typical of warp-knitted non-crimp fabrics. In particular, *weftYarn4* incorporates periodically compacted segments introduced using matrix-soluble functional threads, forming the TFC-integrated architecture. Compaction increases the local fiber volume fraction within these segments while creating adjacent regions with elevated void space. This spatial redistribution of porosity promotes enhanced effective in-plane permeability during the early stage of infusion. Upon thermal activation, the functional threads dissolve and the compacted yarns relax toward their original diameter; consequently, the fiber packing after impregnation and prior to complete crosslinking closely approaches that of the reference architecture.

A representative functional thread is Grilon® KE 60, a thermally activated polyamide developed by EMS Chemie AG. The structural yarn system comprises weft and warp yarns supplied by P-D Glasseiden GmbH, and stitch yarns sourced from Textilglas Technologie GmbH. Two fabric variants (Reference and Modified) with dimensions of 120 × 240 × 1.5 mm were manufactured and are summarized in Table 1. For the 120 mm width, this configuration results in a total of 33 weft tows, of which every fourth weft tow in the corresponding layer is compacted in the Modified BX-NCFs.

2.2 In-Plane Permeability Tests and Flow-Front Tracking

To quantify the influence of weft-direction TFC insertion on the mold-filling rate in resin transfer molding (RTM), one-dimensional in-plane permeability tests were conducted following ISO 4410 to monitor the global flow-front progression (Fig. 2). The preform was placed on a rigid steel base and covered with

Table 1. Specifications and structural comparison of reference and TFC-integrated biaxial glass fabrics.

Biaxial fabric	Manufacturer	Type	Fineness (tex)
Weft Yarn	P-D Glasseiden GmbH	EC 2400	2400
Warp Yarn	P-D Glasseiden GmbH	EC 2400	2400
Stitch Yarn	Textilglas Technologie KG	EC9 34x3	200
	Reference	Modified	
Designed			
Manufactured		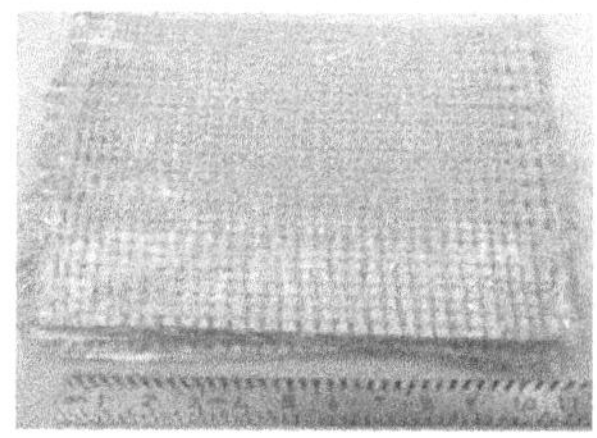	

a transparent glass plate to enable *in situ* visualization of the advancing flow front using a high-resolution camera (Resolution 2560 × 1920 pixels, frame rate up to 3 fps). Spacer plates were used to define the cavity height and control the fiber volume fraction. The edges of the textile were sealed with gelatin to prevent race-tracking effects [20]. A pressure pot injected the test fluid under a constant injection pressure in the range of 0.5–6 bar. The injection inlet and vent outlet are indicated by red arrows in the central panel of Fig. 2. Dimensional stability of the setup was ensured using stiffening crossbeam plates to prevent mold deformation causing irregular permeability distribution. Silicone oil (WACKER® AK 100) was used as the test fluid, with a dynamic viscosity of approximately 0.097 Pa·s and a density of 970 kg/m^3 at a temperature of 25 °C [21].

A custom MATLAB script, termed *Permeability Master* (PM), was developed to automate the image analysis. After positioning the BX-NCF specimen in the test region (indicated by the green box in the central panel of Fig. 2), front-edge pixels are identified at each time step within a predefined region of interest (purple box). These pixels are highlighted in orange and fitted with a straight line (green line) to determine the instantaneous flow-front orientation. As shown in Fig. 2, the flow front is inhomogeneous in the direction perpendicular to the main flow, likely due to the use of a point inlet rather than a planar inlet, as well as local flow hindrance near the specimen edges caused by the gelatin seal. To account for these effects, the global flow front is determined using a width-averaged approach. The width-averaged flow-front position, x, is computed across the textile width and indicated by a white dashed line. An example of real-time flow-front detection for BX-NCFs is shown in the upper

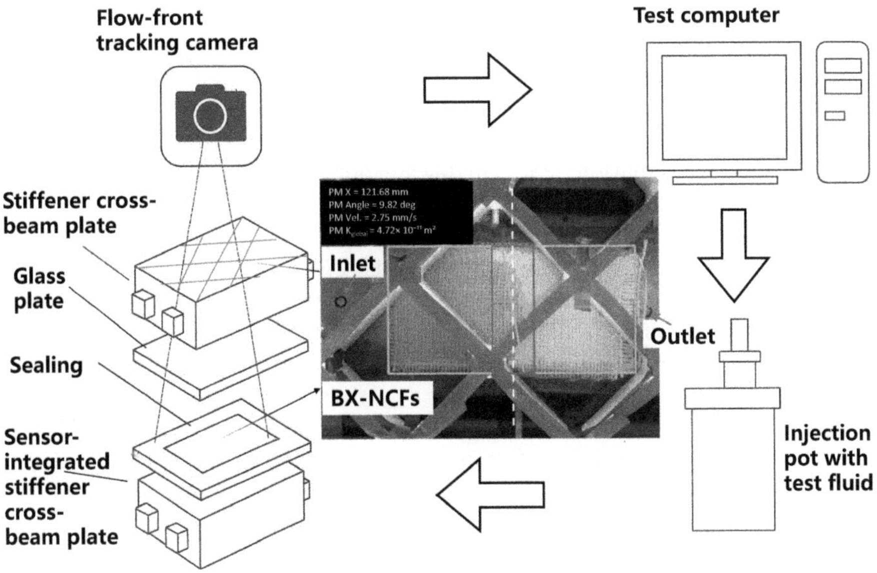

Fig. 2. Experimental setup for in-plane permeability tests and flow-front detection using PM.

left corner of the image, yielding a flow-front position of $x = 121.68$ mm, a flow-front angle of 9.82°. A width-averaged instantaneous flow-front velocity of $2.75\,\mathrm{mm\,s^{-1}}$ was obtained from the flow-front tracking images. However, due to unstable flow behavior near the inlet during the initial filling stage, the inlet velocity applied in the numerical simulations was defined as the time-averaged value of $7.53\,\mathrm{mm\,s^{-1}}$ over the first 0.1 % of the total filling time. The characteristic in-plane permeability of the tested reinforcement is $K_{\mathrm{global}} = 4.72 \times 10^{-11}\,\mathrm{m}^2$, corresponding to a viscous resistance coefficient of $2.12 \times 10^{10}\,\mathrm{m}^{-2}$. This workflow enables continuous tracking of flow-front progression and quantitative evaluation of the transient infiltration velocity [22].

3 Multiphase Flow Modeling in TFC-Integrated BX-NCFs

This section introduces a dual-domain porous–free-flow modeling approach for capturing the dual-scale flow behavior of BX-NCFs and presents a Volume of Fluid (VOF)–based framework with physics-based simplifications.

3.1 Dual-Domain Modeling of Dual-Scale Flow in BX-NCFs

Modeling approaches for resin infiltration in fibrous reinforcements can be broadly classified into pore-scale CFD models, homogenized porous-media models, dual-scale porous-media models, and pore-network models, reflecting different trade-offs between geometric resolution and computational cost [23]. Pore-scale CFD models resolve flow around individual fibers with high fidelity but at high computational cost [24]. Homogenized porous-media models describe the preform as an equivalent continuum with averaged properties and are suitable for macroscopic simulation [25]. Dual-scale porous-media models incorporate multiple length scales to improve accuracy while retaining computational efficiency [26]. Pore-network models represent the pore space as a discrete network, enabling efficient pore-scale transport analysis [27,28].

In BX-NCFs, the textile architecture gives rise to multi-scale flow behavior during mold filling. As illustrated in Fig. 3, resin simultaneously advances through inter-roving macro-channels (characteristic spacing d_1) and intra-roving micro-scale pores (characteristic spacing d_2), giving rise to the *dual-scale effect* that governs flow-front evolution. Owing to their ability to capture this scale interaction while maintaining reasonable computational efficiency, dual-scale porous-media models are therefore adopted to simulate flow behavior and evaluate permeability during fluid infiltration of BX-NCFs in this study.

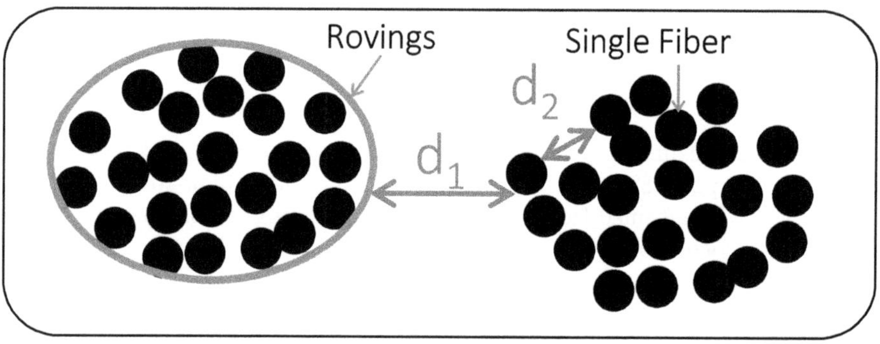

Fig. 3. Schematic illustration of the dual-scale flow behavior in BX-NCFs.

To further account for the strong spatial heterogeneity between inter-roving flow channels and intra-roving porous regions, a dual-domain porous–free-flow modeling approach is adopted, as widely reported in the literature [29]. In the inter-roving regions, where the fluid behaves as a free flow, the motion is governed by the incompressible Navier–Stokes equations [30–32],

$$\nabla \cdot \mathbf{U} = 0, \tag{1}$$

$$\rho \left(\frac{\partial \mathbf{U}}{\partial t} + \mathbf{U} \cdot \nabla \mathbf{U} \right) = -\nabla p + \mu \nabla^2 \mathbf{U}, \tag{2}$$

where $\mathbf{U}$ denotes the velocity vector, p the pressure, ρ the fluid density, and μ the dynamic viscosity.

Within the porous regions corresponding to the fiber rovings, the influence of the textile reinforcement is incorporated through Darcy–Forchheimer resistance terms introduced as momentum source terms [33],

$$-\nabla p = \frac{\mu}{K}\mathbf{U} + C_F \rho |\mathbf{U}| \mathbf{U}, \tag{3}$$

where K is the permeability and C_F is the inertial resistance coefficient. This volume-averaged formulation accounts for both viscous and inertial losses induced by the fibrous microstructure. In the numerical implementation, the Darcy contribution is introduced as a momentum sink term in the Navier–Stokes equations. The viscous resistance source term is expressed as

$$\mathbf{S}_{\text{viscous}} = -\mu K^{-1}\mathbf{v},$$

where μ is the dynamic viscosity, K is the permeability, and $\mathbf{v}$ denotes the superficial velocity. This formulation corresponds to the porous media model implemented in ANSYS Fluent and allows direct specification of the inverse permeability $1/K$ as the viscous resistance coefficient.

3.2 VOF for Multiphase Flow-Front Tracking

Several numerical approaches have been proposed to simulate multiphase flow and flow-front evolution during resin infiltration in fibrous reinforcements. Darcy-based process models predict macroscopic filling but do not explicitly resolve the resin–air interface [34]. Level-set or saturation formulations can represent phase distributions but may require interface reconstruction or face mass-conservation issues [35]. Fully pore-resolved CFD can track the interface accurately but is typically limited to small domains due to computational cost [23].

To capture the advancing resin–air interface with strict mass conservation at the process scale, the Volume of Fluid (VOF) method is employed [36,37]. In the VOF formulation, a single set of governing equations is solved for both phases, and the phase distribution is described by the transport of the resin volume fraction f, which satisfies the volume-fraction conservation equation [36]:

$$\frac{\partial(\rho\phi f)}{\partial t} + \nabla \cdot (\rho \mathbf{U} f) = 0, \tag{4}$$

where ϕ denotes the local porosity. The effective mixture properties are evaluated using volume-fraction-weighted averaging [37]:

$$\rho = f\rho_{\text{resin}} + (1-f)\rho_{\text{air}}, \tag{5}$$

$$\mu = f\mu_{\text{resin}} + (1-f)\mu_{\text{air}}. \tag{6}$$

This formulation enables explicit and mass-conservative tracking of the advancing flow front during resin infiltration.

3.3 Physics-Based Model Simplification

Figure 4 outlines the workflow for extracting and simplifying textile properties for mold-filling simulations. Fiber waviness parameters were derived from X-ray computed tomography (CT) data and subsequently reconstructed using TexGen. Local porosity was quantified from optical micrographs using *ImageJ* [38]. The micrographs were acquired with a spatial resolution of 250 μm and calibrated using a reference length scale to ensure accurate pixel-to-length conversion. Porosity was computed by grayscale thresholding and binary segmentation, and the associated measurement uncertainty was estimated from repeated analyses of representative regions. As illustrated in the lower-left corner of Fig. 4, analysis of a representative micrograph yields a local porosity value of 0.4.

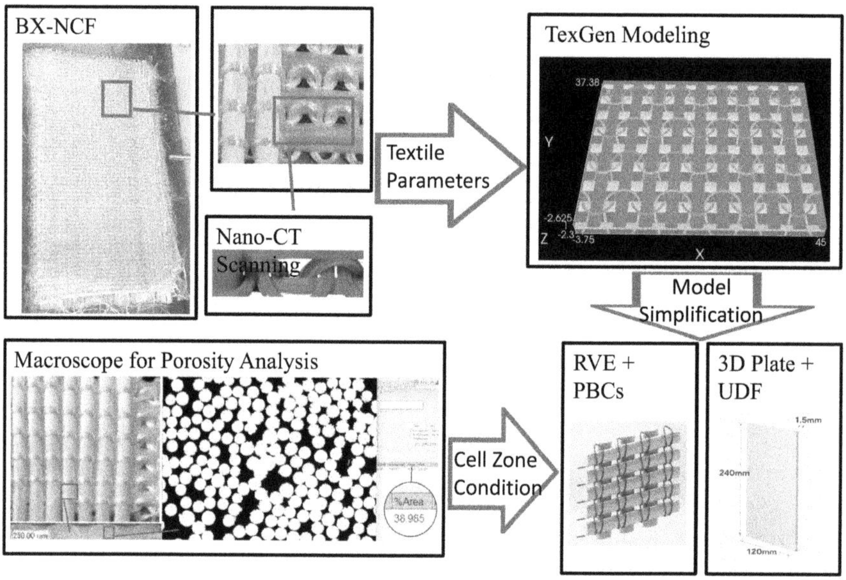

Fig. 4. Workflow for BX-NCF model simplification for mold-filling simulation.

To reduce computational cost while maintaining sufficient accuracy in evaluating the influence of textile architecture, two modeling simplifications are employed, as summarized in the lower-right corner of Fig. 4: (i) a Representative Volume Element (RVE) model with periodic boundary conditions (PBCs) [37,39], and (ii) a sample-scale three-dimensional plate model incorporating a User-Defined Function (UDF)-based porosity assignment [37,40]. Together, these two approaches allow the influence of textile architecture to be analyzed across multiple length scales, from local microstructural effects to sample-scale flow-front evolution.

RVE with PBCs. Figure 5(a) shows the geometry of an RVE consisting of four warp and four weft rovings arranged in a two-layer configuration, while the corresponding finite element discretization is illustrated in Fig. 5(b). The rovings are modeled with elliptical cross-sections defined by semi-axes $a = 0.375$ mm and $b = 1.3$ mm and are arranged at regular in-plane intervals of 1.3 mm. The fourth weft roving is compacted, and its elliptical cross-section is modified such that the semi-axis b is reduced from 1.3 mm to 0.975 mm. The overall dimensions of the RVE are indicated in the figure. The RVE is discretized using structured 8-node hexahedral elements, comprising a total of 4.05×10^6 elements, with mesh refinement applied in the roving regions. A mesh convergence study confirms that the predicted response is independent of the element size. This modeling approach is computationally efficient and well suited for parametric studies of microstructural effects; however, it may not accurately predict responses governed by mesoscale heterogeneity or edge effects present in real textile composites [41].

(a) (b)

Fig. 5. RVE-based geometry used for numerical simulation: (a) BX-NCF Representative Volume Element and (b) hexahedral computational mesh.

Sample-Scale 3D Plate Model with UDF. To enable direct comparison with experimental observations while maintaining computational efficiency, a three-dimensional sample-scale plate model combined with a UDF-based porosity assignment is adopted. As shown in Fig. 6 (left), the plate domain is discretized using a structured hexahedral mesh. For model simplification, the detailed fiber geometry is not explicitly resolved. Instead, the influence of the textile architecture is represented through a spatially varying porosity field implemented via a UDF. This porosity field captures the mesoscale heterogeneity of the BX-NCF architecture, as schematically illustrated in Fig. 6 (right), and enables direct comparison between the macroscopic evolution of the global flow front and experimental measurements.

As shown in Fig. 7, the UDF captures the periodic architecture of the BX-NCF along the warp and weft directions. Fiber tows are idealized as rectangular cross-sections of 2.6 mm × 0.75 mm and arranged at regular in-plane intervals

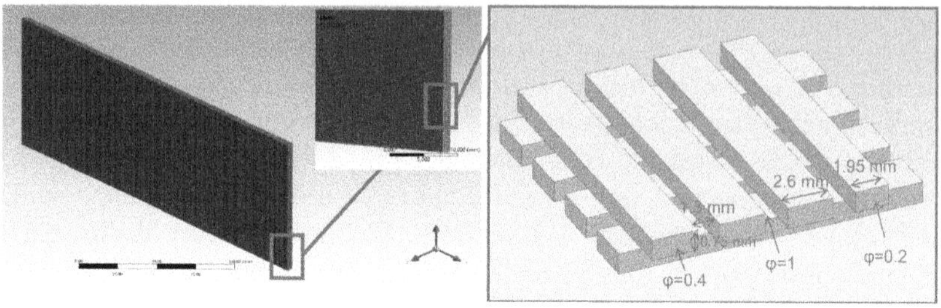

Fig. 6. 3D plate model with hexahedral discretization (left) and schematic of the UDF-based porosity distribution representing mesoscale textile heterogeneity (right).

of 1.3 mm. Regions occupied by fiber tows are assigned a porosity of 0.4 based on image-based analysis, while inter-tow channels are treated as voids with a porosity of 1.0. For regions where the z-coordinate exceeds 0.75 mm, porosity is assigned following a fiberwise description along the y-direction (weft). In the modified BX-NCF configuration, every fourth weft tow in this layer is modeled as a compacted tow with a reduced rectangular cross-section of 1.95 mm $\times$ 0.75 mm. Assuming incompressible fibers and one-dimensional compaction, this geometric reduction implies a decrease in local porosity from 0.4 to 0.2. Accordingly, a porosity value of 0.2 is assigned at these modified locations to reflect the increased local fiber volume fraction.

4 Results Discussion

This section investigates the one-dimensional flow-front evolution in BX-NCFs using a numerical two-phase flow model, and compares the resin flow rate of reference and TFC-modified fabrics through combined experimental and numerical results to assess flow-front behavior.

4.1 One-Dimensional Flow-Front Development in BX-NCFs

The one-dimensional flow-front evolution observed experimentally was reproduced numerically using ANSYS Fluent. A dual-domain porous–free-flow modeling strategy combined with a Volume of Fluid (VOF) formulation was employed. Within this framework, two physics-based model simplifications were adopted, enabling the influence of textile architecture to be analyzed across multiple length scales, as summarized in Sect. 3.3. The physical properties of the test fluid (Sect. 2.2) and the textile architecture parameters (Sect. 3.3) were obtained experimentally and applied consistently in both numerical models.

RVE model with PBCs. Figure 8 illustrates the simulated flow-front development within a representative volume element (RVE). The RVE geometry, shown in Fig. 5(a), and its discretization using structured hexahedral elements (Fig. 5(b))

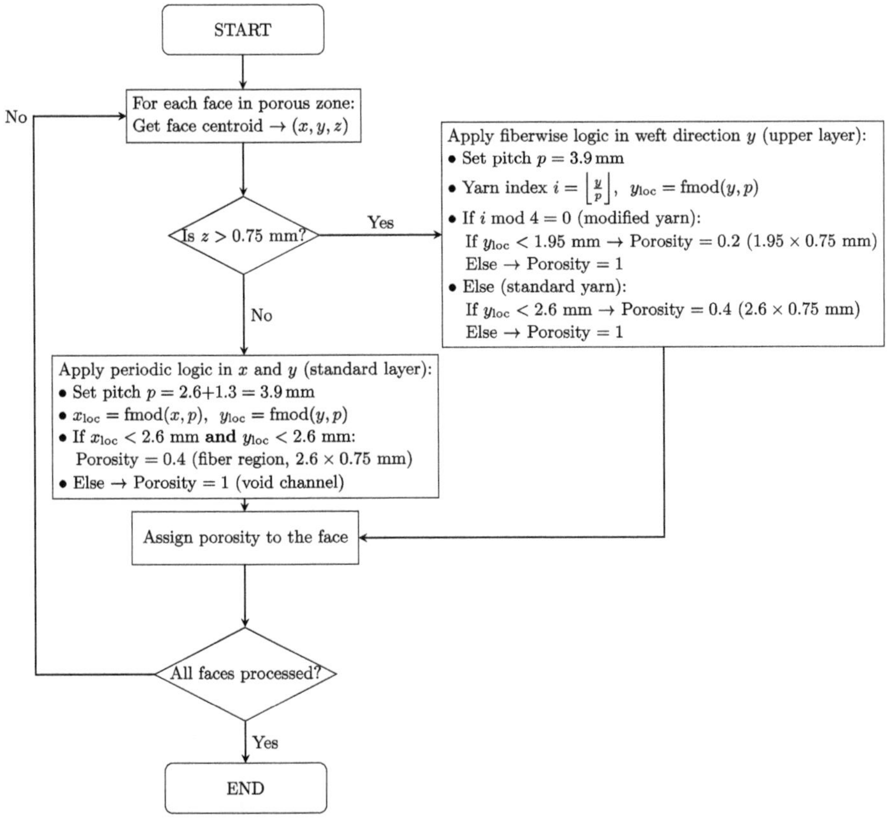

Fig. 7. Flowchart of the porosity assignment algorithm used in the UDF implementation.

are introduced in Sect. 3.3. Periodic boundary conditions were applied on all in-plane faces, including the flow direction, to enforce geometric and kinematic periodicity of the textile architecture. Flow was driven by imposing a prescribed pressure gradient across the periodic faces in the flow direction, corresponding to the experimentally measured superficial velocity. The boundaries in the thickness direction were treated as no-slip walls to represent mold-wall confinement.

Figure 8 presents top- and bottom-view visualizations of the RVE at the same simulation time step, while the central image shows the experimentally observed local flow front in a BX-NCF specimen comprising three rovings. In the top layer (Fig. 8(a)), where fibers are aligned with the flow direction, the test fluid advances preferentially through macro-channels between rovings, resulting in a non-uniform, finger-like flow front, as indicated by the white dashed line. This behavior is consistent with the experimentally observed flow-front shape, marked by the dashed green line in the central panel of Fig. 8(a). In contrast, the bottom layer (Fig. 8(b)), with fibers oriented perpendicular to the flow direction,

Scale	(a) Top layer	(b) Bottom layer

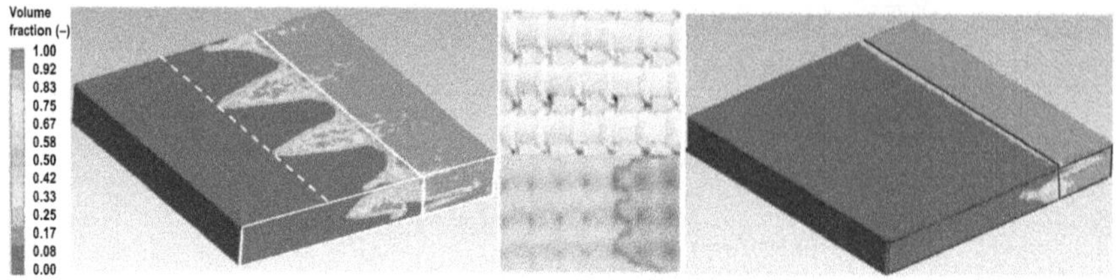

Fig. 8. Simulated flow-front behavior in the top and bottom layers of the BX-NCF architecture using the RVE model.

exhibits a more uniform flow-front progression due to increased flow resistance. The simulated flow front in the RVE, indicated by the solid black line, represents the saturation interface between fully saturated and unsaturated regions and is highlighted by the solid white line in the top layer. These observations illustrate the dual-scale flow mechanism, in which macro-channel-dominated transport coexists with fiber-bundle infiltration, and demonstrate the potential of flow-direction-oriented TFC modification to promote preferential in-plane flow.

3D plate model with UDF. In addition to the RVE simulations, a sample-sized simplified three-dimensional plate model combined with a User-Defined Function (UDF) was employed to capture macroscopic flow-front development. As described in Sect. 3.3, the geometry, shown in the lower-right corner of Fig. 4, was discretized using structured hexahedral elements with a uniform element size of 0.3 mm, resulting in approximately 1.6×10^6 elements (Fig. 6 (left)) to ensure numerical stability. Spatially varying porosity fields were prescribed via a user-defined function (UDF) (Fig. 7) to represent the heterogeneous textile architecture. Velocity-driven boundary conditions were applied, with a uniform inlet velocity of 7.53 mm/s corresponding to the experimental conditions described in Sect. 2.2, and a gauge pressure outlet of 0 Pa to represent a vented outlet open to the atmosphere. All remaining surfaces were treated as no-slip walls. Within the VOF framework, pure oil (volume fraction = 1) was imposed at the inlet, whereas the outlet was defined as a vent with air specified as the backflow phase to allow displacement of air during mold filling. Figure 9 shows the flow-front development at a simulation time of approximately 6.3 s.

Consistent with the RVE results shown in Fig. 8, the top layer of the three-dimensional plate model exhibits a finger-like flow-front morphology (Fig. 9(a)), whereas the bottom layer shows a more uniform flow-front progression (Fig. 9(b)). However, due to the UDF-based homogenization of the textile architecture, in which fiber cross-sections are approximated as rectangular rather than oval, the flow front in the top layer appears more pronounced and non-uniform. This leads to elongated streaks extending ahead of the fully saturated region. Such streaks are difficult to distinctly capture in the experiments;

Scale	(a) Top layer	(b) Bottom layer

Fig. 9. Simulated flow-front development in the top and bottom layers of the BX-NCF using the three-dimensional plate model.

instead, the saturation interface is primarily observed as a straight line oriented perpendicular to the main flow direction. This interface, which is also clearly visible in the bottom layer, can be readily identified in the experiments due to the strong optical contrast between saturated and unsaturated regions.

4.2 Flow Front Evolution in Reference and Modified Textiles

To validate the numerical simulations against the experimental observations, a dual-domain porous–free-flow modeling strategy was employed using the three-dimensional plate model with UDF-based porosity assignment. Figure 10 compares the temporal evolution of the flow front for two textile configurations: the **reference** BX-NCF (Fig. 10(a)) and the **modified** BX-NCF (Fig. 10(b)). Each row in Fig. 10 presents the flow-front progression at three selected time steps, with the experimental observations shown in the upper half of each image and the corresponding simulation results displayed in the lower half. In the modified configuration, TFCs are integrated periodically within the weft yarns, resulting in locally reduced fiber cross-sections and increased porosity, which are explicitly represented in the numerical model through the UDF-based porosity definition. Consistent with the numerical simulation observed in Fig. 9, the top layer of the three-dimensional plate model exhibits elongated streak-like flow-front features. However, the saturation interface is predominantly characterized by a straight front oriented perpendicular to the main flow direction, which agrees well with the experimentally extracted global flow front indicated by the white dashed line.

Figure 11 illustrates the evolution of the flow front for both the **Reference** and **Modified** configurations as a function of filling time. A strong consistency is observed between the experimental measurements and numerical predictions, with coefficients of determination of $R^2 = 0.985$ for the Reference case and $R^2 = 0.993$ for the Modified case. In addition to this high level of agreement, the Modified configuration exhibits a significantly reduced filling time compared to the Reference case, decreasing from 40.6 s to 35.5 s. This corresponds to a

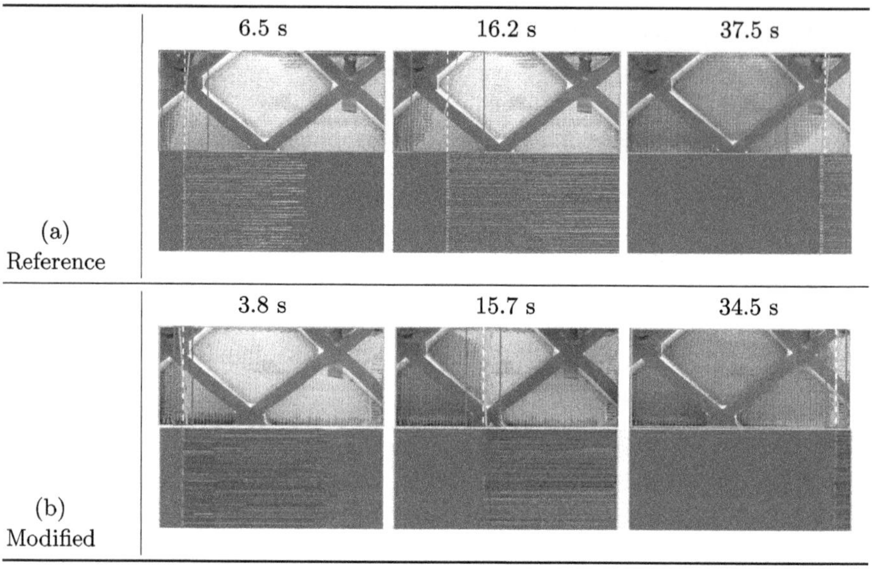

Fig. 10. Flow front progression during infusion for (a) Reference and (b) Modified textiles at selected time points.

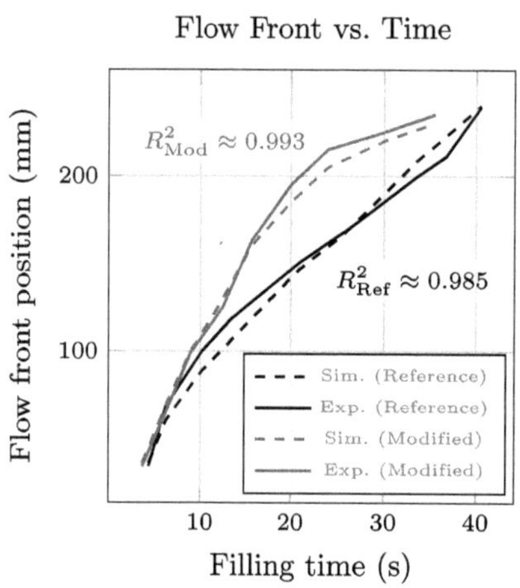

(a) Flow front vs. filling time

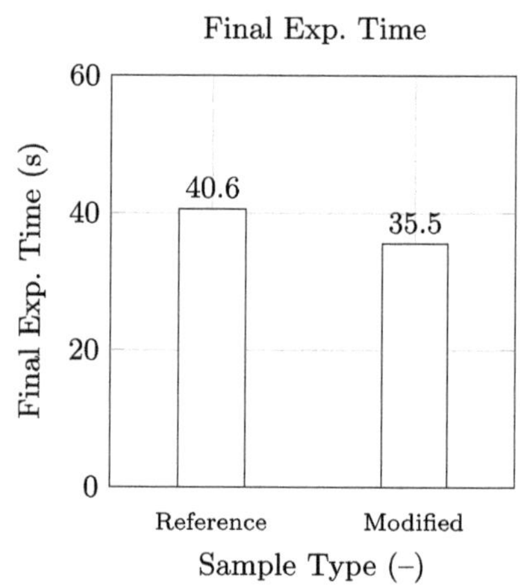

(b) Final experimental filling time

Fig. 11. Experimental and numerical comparison of flow behavior in Reference and Modified BX-NCFs: (a) Flow front position versus filling time; (b) Final filling time.

time saving of 5.1 s, equivalent to a 12.6 % improvement in mold-filling efficiency. These results reflect the enhanced flow characteristics induced by the TFC modification and demonstrate both improved process efficiency and strong predictive capability of the numerical model.

5 Conclusion

This study demonstrated the effectiveness of integrating TFCs into BX-NCFs to enhance the mold-filling rate in LCM processes. By leveraging the dual-scale flow behavior intrinsic to the textile architecture, the incorporation of TFCs enabled a substantial improvement in in-plane resin distribution. Experimental investigations and numerical simulations consistently showed that the modified fabric architecture led to a 12.6 % reduction in filling time compared to the conventional configuration, without compromising final structural integrity. The strong agreement between simulation and experimental results, with coefficients of determination of $R^2 = 0.985$ for the reference structure and $R^2 = 0.993$ for the modified structure, validated the accuracy and reliability of the numerical model for predicting resin infiltration in complex textile assemblies. This improvement is specific to the studied laminate architecture and applies only to the flow direction along the modified textile under the tested conditions; extension to other laminate thicknesses and flow directions will be investigated in future work. Overall, these results underscore the potential of structurally engineered reinforcements, such as TFC-enhanced BX-NCFs, to address flow-related challenges in thick laminates, paving the way for more efficient and dependable composite manufacturing processes.

Acknowledgements. The author gratefully acknowledges the project partners at TU Dresden for their valuable support and insightful discussions throughout the development of this work. Furthermore, financial support from the IGF project (22908 BG/2) and the Industrielle Gemeinschaftsforschung (IGF) is gratefully acknowledged. Without this support, the present work would not have been possible.

References

1. Abliz, D., Ziegmann, G.: Liquid composite molding processes. In: M. Sinapius, G. Ziegmann (eds.) Acting Principles of Nano-Scaled Matrix Additives for Composite Structures, Research Topics in Aerospace, pp. 79–88. Springer, Cham (2021). https://doi.org/10.1007/978-3-030-68523-2_5
2. Abliz, D., Jürgens, T., Artys, T., Ziegmann, G.: Cure kinetics and rheology modelling of boehmite (AlOOH) nanoparticle modified epoxy resin systems. Thermochim. Acta **669**, 30–39 (2018). https://doi.org/10.1016/j.tca.2018.06.017
3. Birkefeld, K.: Virtuelle optimierung von geflecht-preforms unter berücksichtigung von fertigungsaspekten, Universität Stuttgart (2013). Ph.D. thesis
4. Flemming, M., Ziegmann, G., Roth, S.: Faserverbundbauweisen: Halbzeuge und Bauweisen, Springer (1996). https://doi.org/10.1007/978-3-642-61432-3
5. Lefevre, D., Comas-Cardona, S., Binetruy, C., Krawczak, P.: Modelling the flow of particle-filled resin through a fibrous preform in liquid composite molding technologies. Compos. Part A **38**(10), 2154–2163 (2007). https://doi.org/10.1016/j.compositesa.2007.06.008
6. Sutherland, L.S., Soares, C.G.: Effect of laminate thickness and of matrix resin on the impact of low fiber-volume, woven roving e-glass composites. Compos. Sci. Technol. **64**(13), 1691–1700 (2004). https://doi.org/10.1016/j.compscitech.2004.01.017

7. Chen, S., Joesbury, A.M., Yu, F., Harper, L.T., Warrior, N.A.: Optimisation of intra-ply stitch removal for improved formability of biaxial non-crimp fabrics. Compos. Part B Eng. **229**, 109,464 (2022). https://doi.org/10.1016/j.compositesb.2021.109464
8. Schäfer, B., Dörr, D., Naouar, N., Wank, J.P., Kärger, L.: Membrane behavior of uni- and bidirectional non-crimp fabrics in off-axis-tension tests. Int. J. Mater. Form. **16**(6), 68 (2023). https://doi.org/10.1007/s12289-023-01792-x
9. Yang, B., You, Y., Leu, M.C., Li, L.: Influence of fabric shear and flow direction on void formation in resin transfer molding. Compos. Part A **72**, 63–73 (2015). https://doi.org/10.1016/j.compositesa.2015.01.008
10. Gascón, L., Codina, R., Sánchez, J.A.: Numerical prediction of saturation in dual-scale fibrous reinforcements. Compos. Part A **75**, 33–42 (2015). https://doi.org/10.1016/j.compositesa.2015.02.003
11. Cherif, C.: Textile Materials for Lightweight Constructions, Springer, Heidelberg (2016). https://doi.org/10.1007/978-3-662-46341-3
12. Trümper, W., Lin, H., Callin, T., Bollengier, Q., Cherif, C., Krzywinski, S.: Recent developments in multi-layer flat knitting technology for waste-free production of complex shaped 3d-reinforcing structures for composites. IOP Conf. Ser.: Mater. Sci. Eng. **141**, 012,015 (2016). https://doi.org/10.1088/1757-899X/141/1/012015
13. Szarski, M.: Instant flow distribution network optimization in liquid composite molding. Int. J. Adv. Manuf. Technol. **124**, 2271–2285 (2023). https://doi.org/10.1007/s10845-022-01990-5
14. May, D., Loeschke, S., Pinter, G., Stapf, M., Schulte, K.: In-plane permeability characterization of engineering textiles based on radial flow experiments: a benchmark exercise. Compos. Part A **121**, 100–114 (2019). https://doi.org/10.1016/j.compositesa.2019.03.006
15. Ameri, E., Lebrun, G., Laperrière, L.: In-plane permeability characterization of a unidirectional flax/paper reinforcement for liquid composite molding processes. Compos. Part A **85**, 52–64 (2016). https://doi.org/10.1016/j.compositesa.2016.03.002
16. Yang, W., Chen, W.: Simulation method for filling stage in liquid composite molding based on a two-phase flow volume of fluid model. Phys. Fluids **36**, 043,110 (2024). https://doi.org/10.1063/5.0172521
17. Jo, H., Bae, S., Hong, H., Kim, W., Kim, S.S.: Prediction of transverse permeability in representative volume elements with closely arranged fibers through the application of delaunay triangulation and electrical-circuit analogy. Compos. Struct. **334**, 117,984 (2024). https://doi.org/10.1016/j.compstruct.2024.117984
18. Naouar, N., Vidal-Sallé, E., Schneider, J., Maire, E., Boisse, P.: 3D composite reinforcement meso-fe analyses based on x-ray computed tomography. Compos. Struct. **132**, 1094–1104 (2015). https://doi.org/10.1016/j.compstruct.2015.07.005
19. Brown, L.P., Long, A.C.: Modeling the geometry of textile reinforcements for composites: Texgen. Presented at the (2021). https://doi.org/10.1016/B978-0-12-819005-0.00008-3 Composite Reinforcements for Optimum Performance, 2 edn
20. International Organization for Standardization: Iso 4410:2023 — test methods for the experimental characterization of in-plane permeability of fibrous reinforcements for liquid composite moulding. Standard (2023). https://www.iso.org/standard/79944.html. Accessed 2025-10-16
21. WACKER Chemie AG: WACKER® AK 100 Silicone Fluid: Technical Data Sheet (2025). https://www.wacker.com. Online resource, last updated 2025-01-22

22. Fauster, E., et al.: Image processing and data evaluation algorithms for reproducible optical in-plane permeability characterization by radial flow experiments. J. Compos. Mater. **53**, 45–63 (2018). https://doi.org/10.1177/0021998318780209
23. Soulaine, C., Tchelepi, H.A.: Micro-continuum approach for pore-scale simulation of subsurface processes. Transp. Porous Media **113**(3), 431–456 (2016). https://doi.org/10.1007/s11242-016-0701-3
24. Icardi, M., Mancini, S., Bear, J., et al.: Pore-scale simulation of fluid flow and solute dispersion in porous media. Phys. Rev. E **90**, 013,032 (2014). https://doi.org/10.1103/PhysRevE.90.013032
25. Nguyen, V.H., Deléglise-Lagardère, M., Park, C.H.: Modeling of resin flow in natural fiber reinforcement for liquid composite molding processes. Compos. Sci. Technol. **113**, 38–45 (2015). https://doi.org/10.1016/j.compscitech.2015.03.016
26. Bancora, S., et al.: Efficient dual-scale flow simulation for resin transfer molding process based on domain skeletonization. Compos. Part A **167**, 107,624 (2023). https://doi.org/10.1016/j.compositesa.2022.107624
27. Thauvin, F., Mohanty, K.K.: Network modeling of non-darcy flow through porous media. Transp. Porous Media **31**, 19–37 (1998). https://doi.org/10.1023/A:1006558926606
28. Xiong, Q., Baychev, T.G., Jivkov, A.P.: Review of pore network modelling of porous media: Experimental characterisations, network constructions and applications to reactive transport. J. Contam. Hydrol. **192**, 101–117 (2016). https://doi.org/10.1016/j.jconhyd.2016.07.002
29. Carlone, P., Rubino, F., Paradiso, V., Tucci, F.: Multi-scale modeling and online monitoring of resin flow through dual-scale textiles in liquid composite molding processes. Int. J. Adv. Manuf. Technol. **96**, 2215–2230 (2018). https://doi.org/10.1007/s00170-018-1703-9
30. Hirsch, C.: Numerical Computation of Internal and External Flows: The Fundamentals of Computational Fluid Dynamics, 2nd edn. Butterworth-Heinemann, Oxford (2007)
31. Versteeg, H.K., Malalasekera, W.: An Introduction to Computational Fluid Dynamics: The Finite, vol. Method, 2nd edn. Pearson Education, Harlow (2007)
32. Ferziger, J.H., Perić, M.: Computational Methods for Fluid Dynamics, 4th edn. Springer, Cham (2019)
33. Whitaker, S.: The Forchheimer equation: a theoretical development. Transp. Porous Media **25**(1), 27–61 (1996). https://doi.org/10.1007/BF00141261
34. Bear, J.: Dynamics of Fluids in Porous Media. Dover Publications, New York (1972)
35. Osher, S., Sethian, J.A.: Fronts propagating with curvature-dependent speed: algorithms based on Hamilton-Jacobi formulations. J. Comput. Phys. **79**(1), 12–49 (1988). https://doi.org/10.1016/0021-9991(88)90002-2
36. Hirt, C.W., Nichols, B.D.: Volume of fluid (vof) method for the dynamics of free boundaries. J. Comput. Phys. **39**, 201–225 (1981). https://doi.org/10.1016/0021-9991(81)90145-5
37. Ansys Inc: ANSYS Fluent Theory Guide, ANSYS Inc (2024). Release
38. Collins, T.J.: Imagej for microscopy. Biotechniques **43**, S25–S30 (2007). https://doi.org/10.2144/000112517
39. Trofimov, A., Ravey, C., Droz, N., Therriault, D., Lévesque, M.: A review on the Representative Volume Element-based multi-scale simulation of 3d woven high performance thermoset composites manufactured using resin transfer molding process. Compos. Part A **169**, 107,499 (2023). https://doi.org/10.1016/j.compositesa.2023.107499

40. Geers, M.G.D., Kouznetsova, V.G., Brekelmans, W.A.M.: Multi-scale computational homogenization: trends and challenges. J. Comput. Appl. Math. **234**(7), 2175–2182 (2010). https://doi.org/10.1016/j.cam.2009.08.077
41. Ostoja-Starzewski, M.: Material spatial randomness: from statistical to representative volume element. Probabilistic Eng. Mech. **21**(2), 112–132 (2006). https://doi.org/10.1016/j.probengmech.2005.07.007

Author Index

N. Merkert and A. Herzog (Eds.): SimScience 2025, CCIS 3027, p. 103, 2026.
https://doi.org/10.1007/978-3-032-29720-4

www.ingramcontent.com/pod-product-compliance
Ingram Content Group UK Ltd.
Pitfield, Milton Keynes, MK11 3LW, UK
UKHW022111180926
475695UK00002B/96

* 9 7 8 3 0 3 2 2 9 7 1 9 8 *